SCIENCE AND HYPOTHESIS

THE UNIVERSITY OF WESTERN ONTARIO SERIES IN PHILOSOPHY OF SCIENCE

A SERIES OF BOOKS

IN PHILOSOPHY OF SCIENCE, METHODOLOGY,

EPISTEMOLOGY, LOGIC, HISTORY OF SCIENCE,

AND RELATED FIELDS

VOLUME 19

LARRY LAUDAN

University of Pittsburgh

SCIENCE AND HYPOTHESIS

Historical Essays on Scientific Methodology

D. REIDEL PUBLISHING COMPANY

DORDRECHT : HOLLAND / BOSTON : U.S.A.

LONDON : ENGLAND

Library of Congress Cataloging in Publication Data

Laudan, Larry.
Science and hypothesis.

(University of Western Ontario series in philosophy of science; v. 19)
Includes bibliographical references and index.
1. Science–Methodology–Addresses, essays, lectures.
I. Title II. Series.
Q175.3.L38 502'.8 81-15423
AACR2

ISBN 978-90-481-8367-8

Published by D. Reidel Publishing Company,
P.O. Box 17, 3300 AA Dordrecht, Holland.

Sold and distributed in the U.S.A. and Canada
by Kluwer Boston Inc.,
190 Old Derby Street, Hingham, MA 02043, U.S.A.

In all other countries, sold and distributed
by Kluwer Academic Publishers Group,
P.O. Box 322, 3300 AH Dordrecht, Holland.

D. Reidel Publishing Company is a member of the Kluwer Group.

To My
Mother and Father

CONTENTS

ACKNOWLEDGEMENTS

The appearance of this book is due in very large measure to the encouragement of my friend Robert Butts. It was he who saw some redeeming features in my early studies on the history of methodology and who has been an invaluable critic of my work ever since. There are, however, numerous others who have contributed to my understanding of the issues addressed here. Among those who deserve special mention are all of the following: Gerd Buchdahl, Adolf Grünbaum, Rachel Laudan, Peter Machamer, Alex Michalos, John Nicholas, Tom Nickles, Ilkka Niiniluoto, Wesley Salmon, Jerome Schneewind, John Strong and Wolfram Swoboda.

Contemporary scholarship depends as much on institutions as on individuals. Generous support from the American Council of Learned Societies, the Royal Society of London, the National Science Foundation and the University of Pittsburgh made it possible to travel to relevant manuscript repositories. Librarians at the Royal Society of London, the University of Aberdeen, the University of Geneva, and the Wren Library at Trinity College (Cambridge) provided important assistance.

I also want to thank the publishers and editors of several journals and anthologies for permission to use material which originally appeared under their imprint. Although such material has been extensively revised prior to inclusion in this volume, the original sources are as follows: 'The Sources of Modern Methodology', in Butts and Hintikka (eds.), *Historical and Philosophical Dimensions of Logic, Methodology, and Philosophy of Science* (D. Reidel Publishing Company, Dordrecht, 1977), pp. 3–19; 'The Clock Metaphor and Probabilism', *Annals of Science* (Taylor and Francis) **22** (1966), pp. 73–104; 'The Nature and Sources of Locke's Views on Hypotheses', *Journal of the History of Ideas* **23** (1967), pp. 211–23; "Ex-Huming Hacking', *Erkenntnis* **13** (1978), pp. 417–35; 'Thomas Reid and the Newtonian Turn of British Methodological Thought', in Butts and Davis (eds.), *The Methodological Heritage of Newton* (Toronto University Press, Toronto, 1970), pp. 103–31, 'The Medium and Its Message', in Cantor and Hodge (eds.), *The Subtler Forms of Matter* (Cambridge University Press, Cambridge, 1981); 'Towards a Reassessment of Comte's "Méthode Positive" ', *Philosophy of Science* **38** (1971), pp. 35–53; 'William Whewell on the Consilience of

Inductions', *The Monist* **55** (1971), pp. 368–91; 'Why was the Logic of Discovery Abandoned?' in T. Nickels (ed.), *Scientific Discovery, Logic & Rationality* (D. Reidel Publishing Company, Dordrecht, 1980), pp. 173–83 'Induction and Probability in the 19th Century', in P. Suppes (ed.), *Logic, Methodology and Philosophy of Science IV* (North Holland: Amsterdam, 1973), pp. 429–38; 'The Methodological Foundations of Mach's Anti-Atomism and their Historical Roots', in Machamer and Turnbull (eds.), *Space and Time: Matter and Motion* (Ohio State, Columbus, 1976), pp. 390–417; and 'Peirce and the Trivialization of the Self-Corrective Thesis', in Giere and Westfall (eds.), *Foundations of the Scientific Methodology in the 19th Century* (Indiana University Press, Bloomington, Ind.), pp. 275–306.

Finally, I own a large debt to Carol Vidanoff's tireless efforts to meet a demanding schedule for the completion of this manuscript, to my daughter's indexical skills, and to my wife's sustaining conviction that it was worth the trouble.

Every theory of knowledge is itself influenced by the form which science takes at the time and from which alone it can obtain its conception of the nature of knowledge. In principle, no doubt, it claims to be the basis of all science but in fact it is determined by the condition of science at any given time.

Karl Mannheim,
Ideology and Utopia (1931)

CHAPTER 1

INTRODUCTION

This book consists of a collection of essays written between 1965 and 1981. Some have been published elsewhere; others appear here for the first time. Although dealing with different figures and different periods, they have a common theme: all are concerned with examining how the method of hypothesis came to be the ruling orthodoxy in the philosophy of science and the quasi-official methodology of the scientific community.

It might have been otherwise. Barely three centuries ago, hypothetico-deduction was in both disfavor and disarray. Numerous rival methods for scientific inquiry – including eliminative and enumerative induction, analogy and derivation from first principles – were widely touted. The method of hypothesis, known since antiquity, found few proponents between 1700 and 1850. During the last century, of course, that ordering has been inverted and – despite an almost universal acknowledgement of its weaknesses – the method of hypothesis (usually under such descriptions as 'hypothetico-deduction' or 'conjectures and refutations') has become the orthodoxy of the 20th century.

Behind the waxing and waning of the method of hypothesis, embedded within the vicissitudes of its fortunes, there is a fascinating story to be told. It is a story that forms an integral part of modern science and its philosophy. This book does not quite manage to tell that story; it is too episodic and too brief to manage to accomplish such a task. But it does explore many of the pieces of that conceptual and historical puzzle. It seeks to identify some of the recurrent themes associated with debates about the method of hypothesis and to examine some of the specific causes for its changing fortunes.

At a very different level, this book suggests by example the necessity for re-thinking some pervasive stereotypes about the nature of the philosophy of science. In the overwhelming majority of scholarly studies, the history of the philosophy of science is regarded simply as a part of the history of philosophy. Indeed, many writers set out to tell the history of the philosophy of science by asking the question: what have the great philosophers said about science? That is an interesting question, to be sure, but it hopelessly obscures one's historical understanding, for over and again the decisive and incisive moves within the philosophy of science have come from thinkers –

whether scientists or philosophers – who were well outside the philosophical mainstream. More specifically, the philosophy of science, as the etymology of the phrase suggests, is tied much more closely to developments *within science itself* than many scholars have recognized. As I seek to show in what follows, several of the twists and turns in the fortunes of the method of hypothesis are unintelligible unless seen within the context, and against the backdrop, of contemporary scientific debates. If this book has a message, beyond recounting one small part of the history of the philosophy of science, it is simply that that we must change the way in which we are accustomed to thinking about the relationship between the philosophy of science and that science of which it is the philosophy. The two have traditionally existed in an historical symbiosis which is deeply at odds with the character of their present-day relationship. This book is an exploration, using history as a probe, of alternative conceptions of what the philosophy of science once was, and of what it might again become.

In bringing these essays (many of which were originally published in rather obscure places) together between two covers, I am acutely aware of the fact that they exhibit many flaws. If I were writing them afresh, there is much that I would change, and even more that I would add by way of refinement and qualification. New scholarly studies, published since some of the following chapters first appeared, have brought to light a number of texts and issues which have an important bearing on the material discussed here. But, that said, I remain persuaded that most of the central theses propounded here are close to the mark and that all of the major interpretations I have offered remain plausible accounts of what transpired.

What needs to be stressed above all else, and this would remain true even if I had extensively re-written these essays, is the highly partial and tentative knowledge we have of even the main themes within the history of the philosophy of science. I am painfully aware of the exploratory and defeasible character of much that follows. But given the state of our historical knowledge of those issues, it seems unlikely that an authoritative treatment of them is within our grasp. We are still not sure how to tell the forest from the trees; these essays are very much more an attempt at forest management than a taxonomy of tree-types.

However, some preliminary 'botanizing' is probably in order so as to indicate where this present essay fits within our general scheme of things. Conceived in the broadest fashion, it deals with the history of the philosophy of science. That discipline has two distinct and relatively disjoint parts. One important part of the philosophy of science – one with which this book is *not* concerned – might be called 'the conceptual foundations of the sciences'. Typical examples would be the philosophy of physics or the philosophy of biology. To engage in conceptual foundations is to explore the ontological and epistemological implications and presuppositions of particular scientific theories. The study of conceptual foundations by its very nature is highly content specific. As scientific theories change so do our views of the foundations of science change dramatically. This part of the philosophy of science is, and always has been, integrally linked to the sciences; to engage in conceptual foundations is simultaneously to do science and philosophy.

The second major branch of the philosophy of science is generally called 'the theory of scientific methodology'. The concern here is to understand how scientific theories *in general* are appraised and validated. Because theories of scientific methodology are designed to adjudicate *between* rival scientific theories, they are not meant to be content specific in the sense in which the conceptual foundations of science clearly are. Unfortunately, however, the relative autonomy of theories of methodology from the assertive claims of the specific sciences has led many practising methodologists and many historians of methodology to conclude that the development of methodology has been, and indeed ought to have been, largely independent of the history of science. Where conceptual foundations is generally regarded as posterior to and parasitic upon what happens in science, the theory of methodology is nowadays seen as logically prior to, and independent of, the sciences. I think this view of the nature of methodology is chronically misleading and seek to show some of its inadequacies in the chapters that follow.

With this distinction in hand, it can be said that this book deals with the history of methodology rather than conceptual foundations. But even within methodology itself there are some crucial distinctions to be drawn. To a first approximation, scientific methodology consists of two distinct components: heuristics and validation. The latter enterprise, far more familiar than the former in our time, requires little explanation. Although philosophers of science repeatedly disagree about the specific principles which should govern the testing and validation of theories, it is widely agreed that some such principles are required and that the mechanics of theory validation are a vital part of methodology. Generally, this part of methodology is *epistemological* in its orientation; its concern is to ascertain under what circumstances we can legitimately regard a theory as true, false, probable, verisimilar or close to the truth. (Principles of testing need not be epistemic; our evaluations might, for instance, be pragmatic, but that option is decidedly outside the contemporary philosophical mainstream.) What customarily results from a theory of validation is a set of rules governing the circumstances when scientific theories can be warrantedly *accepted* or *rejected*. Accounts of validation may be either *comparative* or *non-comparative*.

The other branch of methodology, heuristics, is more difficult to characterize precisely. One useful way of defining heuristics is to say that it is concerned with identifying the strategies and tactics that will accelerate the pace of scientific advance. To ask how we can broaden the range of viable theories about a particular domain, or to ask what rules of thumb can assist in the discovery of new theories is to raise heuristic rather than validational questions. Where proponents of theories of validation tend to be preoccupied with truth, falsity and epistemic warrants, the vernacular of heuristic theorists tends to be laced with terms like 'scientific progress' and 'the growth of knowledge'.

These two branches of methodology, although decidedly different in orientation, are not rivals between which one has to choose. It is true that some advocates of validation see no hope for heuristics, and that certain heuristically-oriented methodologists object to the 'foundationalism' implied by theories of validation. But most philosophers of science, both past and present, see heuristics and validation as complementary and equally central parts of a comprehensive theory of scientific methodology.

In seeking, as this volume does, to talk about the history of methodology, our concerns will be simultaneously with heuristic and validational matters. Indeed, one of the most interesting parts of the story deals with the manner in which questions of heuristics and questions of validation gradually came to be distinguished from one another. Just as earlier methodologists tended to confuse epistemic and pragmatic issues, they were also apt to conflate validational and heuristic questions. (Indeed, failure to honor these distinctions continues to create much mischief even in contemporary philosophy of science.)

The method of hypothesis, whose history is the immediate focus of this book, has played a large role in both heuristic and validational discussions of scientific methodology. At least from the time of Bacon onwards, hypotheses and conjectures have been seen to have at least a heuristic role in promoting the advancement of science. Even the many thinkers who would grant no place to the method of hypothesis in the theory of validation have been willing to accord that method a useful heuristic status.

Insofar as there have been disagreements about the method of hypothesis – and I believe those disagreements have perennially been at the very heart of the philosophy of science – they have focussed chiefly on the epistemological or validational significance of that method. Simply put, the problem of hypothesis is this: if we have an hypothesis (or theory) all of whose thus far examined consequences are true, then what – if anything – can we warrantedly infer about the truth or likelihood or verisimilitude or well-testedness of the hypothesis? The easy answer ('nothing') is associated with those who believe that the method of hypothesis is bogus. But if the answer is positive, that is, if the 'confirming' instances of an hypothesis do constitute evidence about its truth status, then what is called for is an epistemic analysis of the kinds of confirming instances which are genuinely rather than spuriously evidential. It is the articulation of such theories of evidence which constitutes a large part of the history of the method of hypothesis.

I said before that the method of hypothesis has played a large role in both theories of validation and in heuristic theories. If there is an emphasis in this book, it tends to be more on the history of validation rather than heuristics. This imbalance reflects both the bias of the age and the bias of the author. Written in a different time, the balance between the history of validation and heuristics would probably be struck rather differently. This is not to apologize for this emphasis; one inevitably brings to the writing of history the preoccupations of one's own time. One's duty is to acknowledge the fact and to insure that it does not badly distort the historical record.

CHAPTER 2

THE SOURCES OF MODERN METHODOLOGY: TWO MODELS OF CHANGE

A GENERAL STATEMENT OF THE PROBLEM

In its most general form, the thesis of this chapter can be succinctly put: we have brought to the writing of the history of methodology certain preconceptions which jointly render it almost impossible to understand the evolution of this subject. These preconceptions concern both the nature of philosophy and the aims of history. Until those conceptions are altered, I venture to claim, the history of methodology will not deserve the serious attention which it would otherwise warrant. I want to talk about this family of related philosophico-historical conceptions, using some specific examples to indicate the problem and to underscore its acuteness.

Most of human kind are what the geologist Charles Lyell would have called 'uniformitarians', in an extended sense of that term. For Lyell himself, of course, to be a uniformitarian was to believe that the prior history of the Earth was in all crucial respects the same as we find it today. In the broader sense in which I shall use the term, a uniformitarian is one who tends to presume that *human* history, whether social or intellectual, has always been pretty much as we find it today. Few modern scholars would admit to being uniformitarians, for they all realize that beliefs, practices, institutions and social conventions can change dramatically through time. Yet despite this realization, the uniformitarian presumption is hard to suppress. In various subtle forms, it still exerts a powerful influence on our vision of the past. Nowhere is this influence more pronounced than in the view of our philosophical past. Especially where the philosophy of science is concerned, there is a tendency to assume that certain features of contemporary philosophy of science have been permanent characteristics of the activity.

This uniformitarianism shows up, not in the crude doctrinal form of suggesting that our predecessors held the same beliefs that we do, but rather in the presumption that the factors that shape this discipline in our time can be taken as substantially the same factors that have always shaped it. More specifically, both philosophers of science and much of the historical scholarship about the philosophy of science presuppose a certain *purist model* of scientific methodology. That model has several prominent features: (1)

it regards the theory of methodology primarily as a philosophical activity and not, except incidentally, as a scientific one; (2) it tends to identify the central themes in the historical evolution of scientific methodology by looking to the writings of the 'great philosophers', paying little more than lip service to the methodological contributions of all but a tiny handful of scientists; (3) it tends to imagine that whenever methodological beliefs have changed – as they often have – these changes must have been grounded in some prior shift in metaphysics and epistemology. The presumption here is that the only considerations relevant to the rational adoption or rejection of methodological positions are 'philosophical' considerations.

There can be little doubt that this purist model has motivated much of the historical scholarship in this field, even much of the best historical scholarship.[1] Consider, for instance, two of the most widely known (and rightly admired) books in this field. Gerd Buchdahl's *Metaphysics and Philosophy of Science*, for all its efforts to stress the importance of physics to the philosophy of science, focusses almost exclusively on the famous philosophers of the 17th and 18th centuries: Descartes, Locke, Berkeley, Hume, Leibniz, and Kant. The story Buchdahl has to tell is a fascinating one, but where the philosophy of science is concerned, it is decidedly skewed in the direction of philosophers, with short shift given to most of the scientists (viz., natural philosophers) of the period. Or consider a slightly older example, the classic *Theories of Scientific Methodology From the Renaissance to the 19th Century* by Blake, Ducasse, and Madden. Apart from a token chapter of Newton, virtually all the figures they discuss enjoyed greater reputations as philosophers than as scientists.

The purist historiography, which sees the history of methodology largely as a philosophical dialogue between applied epistemologists, is quite clearly a correct description of the contemporary state of the art. By and large, the sophisticated discussion of methodological issues in our time is carried on by philosophers and within philosophy departments. The major theorists of scientific methodology in the last fifty years – the likes of Carnap, Popper, Nagel, and Hempel – have been much closer, institutionally and intellectually, to philosophy than to science. Literally no scientists of the last half century would figure on any serious list of eminent theorists of methodology. The purist model described above perfectly captures the insularity of contemporary methodology. But, and this is where the purist model goes badly astray, it has not always been so. Herschel, Bernard, Mach, Helmholtz, Poincaré, and Duhem – to name just a few – were prominent scientists whose contribution to methodology far eclipsed that of any of their philosophical contemporaries.

The fact of the matter is that philosophy of science has traditionally stood in a very different relation both to science and to philosophy from the one now familiar to us. This contrast makes nonsense of the effort to explain the historical evolution of methodology in the same terms in which we would discuss its very recent past. I shall claim that the simplistic projection of certain features of contemporary philosophy of science into the past has rendered it impossible to understand or explain many of the signal developments in scientific epistemology, developments which should be the central foci of the historian. I shall claim, further, that until we recognize that earlier philosophers of science had explanatory ambitions and motivations which were, on the whole, very different from those of their modern-day counterparts, then historical understanding – at even the most superficial level – will elude us. (I should conjoin a caveat here. Purist historiography is not mistaken in its assumption that philosophy has been a fertile source for methodological ideas; that goes without saying. Where the purist approach becomes pernicious is in its emphasis on the exclusivity and primacy of the philosophical roots of methodology.)

My tactics, in seeking to make these claims plausible, will be these: I shall begin by putting forward several general theses about the historical character of philosophy of science; these are meant briefly to outline one possible alternative to the purist's historiography of methodology; I shall then turn to discuss at length a specific historical example which will, I hope, serve as a test case for assessing the relative historical fecundity of the two historiographic models. Later chapters will provide further opportunities to put this rival model to the test.

Looking first at the general theses, I would suggest a pragmatic, symbiotic model for the history of methodology. Its central tenets would be these:

(1) That the original and influential contributions to the development of methodology have come as much from working scientists as from philosophers (insofar as any clear historical distinction can be clearly drawn between scientists and philosophers).

(2) That the epistemological theories of an epoch have frequently been parasitic upon the philosophies of science of that epoch rather than vice versa, i.e., that the methodological ideas of working scientists have often been the source of major epistemological theories.

(3) That the traditional role for the philosopher of science has been predominately descriptive, explicative and legitimative, rather than prescriptive and normative; his avowed aim has been to make explicit

what is already implicit in the best scientific examples, not to reform the best extant scientific practices.

(4) That the reception of methodological doctrines, even by the 'great' philosophers, has been determined more by the capacity of those doctrines to legitimate a preferred scientific theory than by their strictly philosophical merits. Correlatively, when methodologies are rejected it is often because of their inability to rationalize what is regarded intuitively as exemplary scientific practice.

Putting it otherwise: new or innovative methodological ideas have generally not emerged, nor have old ones been abandoned, as the result of an internal, dialectical counterpoint between rival philosophical positions or schools; neither is it the case that the waxing and waning of methodological doctrines can be related neatly to their epistemic credentials. Rather, it is shifting *scientific* beliefs which have been chiefly responsible for the major doctrinal shifts within the philosophy of science.

Much more can, and in the course of this book will, be said about this pragmatic model of methodology. But these remarks should be sufficient for us to begin to test the historical resources of the two models against some important episodes in the history of the philosophy of science.

THE METHOD OF HYPOTHESIS DURING THE ENLIGHTENMENT

It has been known for a long time that the beliefs of philosophers of science about the nature of scientific inference underwent a profound shift between the time of Newton and Mill on a number of fronts. Probably most prominent here were the fortunes of the hypothetico-deductive method (or, as it was less clumsily called at the time, 'the method of hypothesis'). Very briefly put, the method of hypothesis amounted to the claim that an hypothesis could be validated by ascertaining whether all of its examined consequences were true. The logical structure of such inferences took the form:

> H entails $C_1, C_2, \ldots, C_n$.
> $C_1, C_2, \ldots, C_j$ *have been tested and are true.*
> Therefore H is probably true.

Frequently espoused in the middle of the 17th century by Descartes, Boyle, Hooke, Huygens, and the Port-Royal logicians, the method of hypothesis fell into disfavor by the 1720s and 1730s. Few scientists and virtually no philosophers of science had any use for hypothetical inference. Knowing full well the fallacy of affirming the consequent and its implications for the untenability of the method of hypothesis, most scientists and epistemologists accepted the Baconian-Newtonian view that the only legitimate method for science was the gradual accumulation of general laws by slow and cautious inductive methods. Virtually every preface to major scientific works in this period included a condemnation of hypotheses and a panegyric for induction. Boerhaave, Musschenbroek, 'sGravesande, Keill, Pemberton, Voltaire, Maclaurin, Priestley, d'Alembert, Euler, and Maupertuis were only a few of the natural philosophers who argued that science could proceed without hypotheses, and without need of that sort of experimental verification of consequences, which had been the hallmark of the hypothetical method since antiquity. As a contemporary noted, "The [natural] philosophers of the present age hold hypotheses in vile esteem".[2] Philosophers of science and epistemologists, were, if anything, even more enthusiastic in their condemnation of hypothetical inference. Whether we look to Reid's *Inquiries*, to Hume's *Treatise*, to Condillac's *Traitê des Systèmes*, to Diderot's *Discours Prêliminarie*, or even to Kant's first *Critique*, the philosopher's refrain is the same: the method of hypothesis is fraught with difficulties; there are alternative methods of scientific inference, generally thought to be inductive and analogical ones, which alone can generate reliable knowledge. As Thomas Reid put it in 1785: "The world has been so long befooled by hypotheses in all parts of philosophy, that it is of the utmost consequence . . . [for] progress in real knowledge to treat them with just contempt . . . "[3] The ardor with which 18th-century epistemologists repudiated the hypothetical method and endorsed the Newtonian inductive one is, in itself, plausible testimony to the impact of scientific archetypes on epistemology. Because that part of the story is documented below in Chapter 6, I shall not discuss that issue here. What interests me, rather, is a slightly later part of the tale; for the self-same method of hypothesis which was so widely condemned by 18th-century epistemologists and philosophers of science was, three generations later, to be resurrected and to displace the very method of induction which the philosophers and scientists of the Enlightenment had set such store by. If, for instance, we jump ahead to the 1830s and 1840s, methodologists such as Comte, Bernard, Herschel, Apelt, Whewell, Dugald Stewart, and even Mill were prepared to concede that the method of hypothesis had a vital role to play in scientific inference. And, with the exception of Mill, all these thinkers were prepared to acknowledge that the method of hypothesis was, in fact, more central to many forms of scientific inquiry than enumerative or eliminative induction.

This about-turn, which effectively constitutes the emergence of philosophy of science as we know it today, is clearly of great historical importance. Explaining why and how the hypothetico-deductive method – which had been roundly condemned in the 18th century – came into prominence again, should presumably constitute one of the core areas of inquiry for the historian of the philosophy of science. Sadly, nothing could be further from the case. To the best of my knowledge, no scholar has been willing to confront this issue head on. Although several have noted that the process occurred, no one has offered a detailed explanation for it. The avoidance of this historical puzzle cannot be a matter of mere indifference. It has been common knowledge for well over a century that the method of hypothesis waned and waxed between 1720 and 1840. One can only conjecture that the puzzle has not been seriously addressed because the confines within which most historians of philosophy have worked do not allow of any cogent solution. Recall that what I have called the purist historiographical model allows one to explain change within the history of methodology only in terms of the emergence of new philosophical doctrines and arguments. In the case at hand, such an explanation is unavailing because the logical and epistemological arguments which Stewart, Herschel, and Whewell give for reviving the method of hypothesis do not differ substantially from the kind of arguments which its 17th-century partisans had articulated. Indeed, the 19th-century criticisms of induction are, on the whole, no more than variations of criticisms of inductive inference which were well-known in both the 17th and 18th centuries. Since few new philosophical arguments can be found which would explain the re-emergence of the method of hypothesis, or the attendant devaluation of enumerative and eliminative induction, the purist historian has no resources on which to draw for piecing together a rational reconstruction of this decisive episode.

If, however, we are prepared to lay aside the purist model, then we are in a position to offer a plausible intellectual account of the changing philosophical attitudes to the method of hypothesis during the period in question. Although we must look outside of philosophy proper to find the answer, the answer itself tells us something very significant about philosophy itself, and about the ways in which philosophical beliefs are shaped by changes in science.

The chief cause for the shift in attitudes of philosophers towards the hypothetico-deductive methodology was a *prior* shift in attitudes on the part of certain scientists towards the method of hypothesis. That latter shift, in turn, was determined by the changing character of physical theory itself, and by the tensions created when new modes of scientific theorizing ran counter to the inductivist orthodoxy prevalent among 18th-century scientist and philosophers of science. I want to spell this process out in some detail.

For some fifty years after the triumph of Newton's *Principia*, both scientists and philosophers sought to draw the appropriate morals from the Newtonian success. As read by his immediate successors, Newton's achievement depended upon the eschewal of hypothetical reasoning and the rigid adherence to inductive generalizations from experimental data. Whether we look to the work of Hales, Boerhaave, or Cotes, we see an effort to construct a purely observational physics, chemistry, and biology whose ontology is immediately relatable to the data of experience.

By the 1740s and 1750s, however, scientists were discovering that many areas of inquiry did not readily lend themselves to such an approach. As a result, a number of scientists – and philosophers – began developing theories which, in the nature of the case, could not conceivably have been arrived at by enumerative induction. Franklin's fluid theory of electricity, the vibratory theory of heat, the Buffonian theory of organic molecules, and phlogiston chemistry are but a small sample of the growing set of theories in the middle of the 18th century, which hypothesized unobservable entities in order to explain observable processes. Among the most controversial of these theories were the chemical and gravitational theories of George LeSage, the neurophysiological theories of David Hartley, and the general matter theory of Roger Boscovich. Although working completely independently of one another, and differing over many substantive questions, these three thinkers – LeSage, Hartley, and Boscovich – have one very important characteristic in common: they quickly came to realize that the types of theories they were promulgating could not possibly be justified within the framework of an inductivist philosophy of science. Each of these thinkers found that his scientific theories, when once publicized, received widespread criticism, not because of their scientific merits or demerits, but rather because of their alleged epistemic and methodological deficiencies.

Against Hartley, it was claimed that his theory about aetherial fluids in the nervous system was but one of many hypotheses, between which only an arbitrary choice could be made.[4] Against Boscovich, it was argued that he could get no direct evidence that the forces around particles were alternately attractive and repulsive at the microscopic distances where contact, cohesion, and chemical change occurred. Against LeSage, critics contended that his theory of 'ultramondane corpuscles' (corpuscles whose motion and impact explained gravitational attraction) could not be inductively inferred from experiment.

Clearly, what confronted all these scientists was a manifest conflict between the accepted canons of scientific inference and the types of theories they were constructing. There was simply no way to reconcile an inductivist methodology and a sensationist epistemology with such highly speculative theories about micro-structure. Their choice was a difficult one: either abandon micro-theorizing altogether (as their inductive critics insisted) or else develop an alternative epistemology and methodology of science which would provide philosophical legitimation for theories which lacked an inductivist warrant. All three in our trio chose the latter alternative. Boscovich insisted that the method of hypothesis is "the method best adapted to physics" and that, in many cases, it is only by means of conjecture followed by verification that "we are enabled to conjecture or divine the path of truth".[5] Hartley, in a lengthy chapter on methodology in his *Observations on Man*, asserted that the methods of induction must be supplemented by various hypothetical methods if we are ever to accelerate the acquisition of knowledge beyond a snail's pace.[6]

The most explicit defense of hypothesis, however, came from Georges LeSage, whose theory had been severely attacked.[7] Euler, for instance, had said, referring to LeSage's physics, that it was better to remain ignorant "que de recourir à des hypothèses si étranges".[8] The French astronomer Bailly had insisted, in good inductivist fashion, that science should limit itself to those "lois qu'elle nous a manifestées",[9] and avoid conjecturing about what we cannot directly observe. There are, LeSage laments, "almost universal prejudices" that hypothetical reasoning from the observed to the unobserved is impossible, and that induction and analogy are the only legitimate routes to truth.[10] He points out that his own theory was being widely dismissed because "mon explication ne peut être qu'une hypothèse".[11]

Confronted with such attacks, LeSage was forced to play the epistemologist. In several later works, but especially in an early treatise on the method of hypothesis, written for the French *Encyclopédie*, LeSage began the counter-attack. In brief, his strategy was two-fold: first, to establish the epistemic credentials of the method of hypothesis; second, to point up a number of weaknesses in the dominant inductive and analogical accounts of methodology. LeSage, in short, agreed with his critics that his theory was indeed an hypothesis; but unlike them, he sought to show that it was none the worse for that.

He grants immediately that the method of hypothesis and subsequent verification can rarely establish the truth of any general conclusions. But then, as he points out, induction and analogy are also inconclusive. What we must aim at in these matters is high probability and LeSage indicates circumstances under which we are entitled to assert well-confirmed hypotheses with confidence. He goes on to point out that the great Isaac Newton, for all his professed inductivism, extensively utilized the method of hypothesis. It is, he says, to hypothesis, "without any element of [induction or] analogy that we . . . owe the great discovery of the three laws which govern the celestial bodies".[12] Generalizing this point, LeSage argued that there is an element of conjecture or hypothesis in every inductive inference which goes beyond its premises, which all except so-called perfect inductions do. He spent much of the next half century trying to resurrect the method of hypothesis, conceived along these lines.

As I shall show in detail in Chapter 8, the evidence is as unambiguous as evidence can be that LeSage's lengthy and persistent espousal of the method of hypothesis was conditioned by his prior scientific commitments and by the epistemic criticism directed against his scientific theories. Similar, if more circumstantial, claims can also be made for Hartley and for Boscovich.

But the story does not end here, for there is the larger question as to how the method of hypothesis, here propounded by a tiny handful of beleagered scientists, worked its way into philosophical orthodoxy. There are a number of pieces to the larger puzzle. Chief among them are these: Jean Senebier, a French philosopher-scientist best known for his work on photo-synthesis, wrote an influential three-volume work in 1802 on scientific method which, following LeSage, endorses the method of hypothesis and accumulates further evidence for the wide-spread use, among the best physical theorists, of the method of hypothesis.[13] Shortly thereafter, Pierre Prevost, founder of the theory of heat exchange, published a posthumous collection of LeSage's essays on hypothesis and himself wrote a book in 1804 on philosophy of science, in which prominence is given to the method of hypothesis.[14] In Scotland, Dugald Stewart – the leading English-language philosopher at the turn of the century – repudiates the trenchant inductivism of his mentor Reid and, after an explicit discussion of LeSage, Hartley, and Boscovich, warmly endorses the method of hypothesis, since "it has probably been this way that most [scientific] discoveries have been made".[15] As Olson has shown, Stewart's discussion of the philosophy of science provided the framework upon which Herschel and Whewell, prominent proponents of the method of hypothesis, drew heavily.[16] (One might add that the scientific commitments of Herschel and Whewell to the wave theory of light also seem to have had much to do with their espousal of the method of hypothesis; a conjecture which I explore at length in Chapter 10.) In the case of each of these writers – from Hartley and LeSage through Prevost and Stewart to Herschel and Whewell – the explicit and over-riding consideration in their endorsement of the hypothetico-deductive method (as against induction or analogy) was that the sciences of their time required such a method. Although familiar philosophical arguments in favor of the method of hypothesis are routinely rehearsed, and a few important innovations are introduced, the primary factor which justifies that method – in the view of its proponents – is the raw fact that it is being used to good effect in the most successful sciences. Even Mill, who would dearly have loved to dispense with the method of hypothesis, felt constrained – given its widespread use in the sciences – to find a logical rationale for it.[17]

CONCLUSION

Hence, it seems impossible to make any historical sense of this episode by looking at scientific epistemology as autonomous and self-contained. Sadly, but predictably given the predominance of the purist model, one looks in vain in the scholarly literature for any rival to this account of the modern emergence of the hypothetico-deductive method. LeSage, Hartley, Lambert, and Prevost are never even mentioned in histories of the philosophy of science. Although Herschel and Whewell are often discussed, no attention is paid to their intellectual ancestry, nor to the scientific interests which shaped their treatment of hypotheses. Instead, the standard literature leads us by the hand from Newton to Hume to Kant to Whewell, with no hint that most of the methodological theorizing on this vital topic was taking place outside the mainstream – or perhaps I should say outside the presumed mainstream – of epistemology. But is the case typical or is it just the exception which proves the proverbial rule?

Upon the answer to that question hangs the fate of the traditional purist historiography of method. I am inclined to think, judging by some of the best recent scholarly studies, that we are dealing with a common phenomenon. Although such studies are still very much in the minority, it is worth considering some of their conclusions: Mittelstrass and others, following the lead of Duhem, have shown that instrumentalism has its historical roots in debates within ancient and renaissance astronomy;[18] Mandelbaum has shown the intimate connections between atomism and the doctrine of primary-secondary qualities;[19] Edelstein and others have shown the connections between Pyrrhonic scepticism and Greek medicine; Buchdahl and Sabra have explored the impact of optics on epistemology in the 17th century.[20] Recent studies by others have shown that it was ultimately the triumph of atomic-molecular physics in the early years of this century that eventually tipped the balance in the philosophical debate between realists and instrumentalists.[21] The historical record leaves little doubt that the historical fortunes of many philosophies of science have been closely intertwined with the fortunes of those scientific theories upon which the philosophies were modelled.

What all these interconnections suggest is that the philosophy of science and large parts of epistemology have traditionally been modelled on, influenced by, and devised to legitimate, certain preferred or priviliged forms of scientific activity. And why, in a sense, should we have imagined otherwise?

In exploring the nature of science or knowledge, it is perfectly natural that methodologists and epistemologists should have selected as their explicanda the best available examples of science or knowledge. The presumption, so familiar in the 20th century, that general methodology can be entirely normative and logically prior to science and thus not parasitic upon any specific examples of knowledge was not a conception cherished by our predecessors. They were generally prepared to concede that the sciences were justificationally prior to the philosophy of science and that the epistemologist's aim was, in Locke's language, to serve not as judge of, but as underlaborer to, his scientific contemporary. Their task, as they perceived it, was not to *prescribe* what methods the scientist should follow; but to *describe* the best methods to be found in existing scientific *praxis*.

But one would be missing the point to imagine that all that is being claimed here is that scientific theories have somehow inspired or motivated methodological doctrines, and that the latter, once invented, acquired a life of their own and had a temporal career entirely independent of the former. We need to realize that scientific theories not only inspire new theories of methodology, they also – in a curious sense – serve to *justify those methodologies*. For instance, the success of Newton's physics was thought to sanction Newton's rules of reasoning; Lyell's geological theory was cited as grounds for accepting methodological uniformitarianism; the kinetic theory of gases and Brownian motion were thought to legitimate epistemological realism; these are but a few examples of a very common phenomenon.

Given this close symbiotic relation between science and its philosophy, it is entirely natural that when changes take place among the theories regarded as archetypically scientific so, too, should the assessment of methodological doctrines shift. And this is really the nub of the whole matter. Confronted by a range of equally convincing or unconvincing philosophies of science, *our predecessors often looked to the sciences of their time as the appropriate laboratory for evaluating competing philosophies*. They insisted that methodological principles should be 'empirically' tested by seeing whether they could be used to legitimate those scientific theories which were taken to be the best available examples of knowledge. If many of us no longer see the situation in that form, we should nonetheless be sufficiently imaginative historically to be mindful of such issues when we talk about the past. Indeed, whatever the relative merits of these two divergent views about the relation of philosophy of science and the sciences, it is clear that the *historian* must be more prepared than he has been to look beyond the narrow confines of the purely *philosophical* if he is to make any sense of the history of methodology.

My thesis thus far has amounted to this: that one cannot understand the history of methodology without looking carefully at the historical evolution of those sciences upon which methodology has traditionally been parasitic. There is an important corollary to this thesis which has already been made explicit but not discussed; to wit, that precisely because the history of methodology is more than just a branch of the history of epistemology, we must abandon the view, presupposed by so much recent scholarship, that the thinkers who loom large in the history of epistemology should *a forteriori* be accorded pride of place in the history of methodology. It simply is false to assume that there is a neat intersection between the two groups.

Let me draw my examples again from the middle and late 18th century. The prominent epistemologists, of course, were Hume and Reid in Britain, Condillac and Condorcet in France, Wolff and Kant in Germany. Few of these thinkers would figure prominently on a list of original and influential methodologists for the same period. If we would learn who contributed most to such problems as the methods of induction, techniques of hypothesis evaluation, the articulation of various experimental and observational methods, the application of probability theory to scientific inference, the evaluation of claims about theoretical entities, and the like, we must look to a very different constellation of figures. We must look to Hartley, LeSage, and Lambert, who between them elaborated the hypothetico-deductive method at a time when the major philosophers of the day had nothing to say in its favor. We must look to Bernoulli, Mendelsohn, Laplace, and d'Alembert for discussions of the logic of probability. We must look to Pierre Prevost, and to Jean Senebier's classic text on philosophy of science to see the most sophisticated treatment of the various experimental methods and their logical foundations.

The very fact that most of these latter figures are virtually unheard of nowadays, at least within the history of methodology, is further testimony to the extent to which we have allowed our scholarly image of the history of the philosophy of science to be warped by the naive subsumption of methodology under epistemology. If, to return the language of my title, we would learn something about the sources of modern methodology, we must look to a different set of issues and a different set of thinkers from those that have customarily monopolized our attention. More importantly, we must recognize the artificiality of the purist model, and be prepared to see the development of methodology outside of a narrowly philosophical context.

NOTES

[1] For a guide to many of these studies, see my 'Theories of Scientific Method from Plato to Mach: A Bibliographic Review', *History of Science* 7 (1968), 1–63.

[2] Benjamin Martin, *A Philosophical Grammar* (London, 1748), p. 19. Similar sentiments are expressed a year later by Condillac; cf. his *Oeuvres* (Paris, 1798), vol. ii, pp. 327 ff.

[3] T. Reid, *Works* (ed. W. Hamilton, 6th ed., Edinburgh, 1863), vol. 1, p. 236. For further discussion of this background, see Chapter 7.

[4] Most contentious in Hartley's system was his effort to provide a neuro-physiological foundation for the Lockean 'associationist' psychology by postulating an aetherial fluid which filled the nerves.

[5] From Boscovich's *De Solis a Lunae Defectibus* (1760). Quoted from, and translated by D. Stewart in his *Collected Works* (ed. by W. Hamilton, Edinburgh, 1854–60), vol. 2, p. 212.

[6] See, especially, Hartley's *Observations on Man: His Frame, His Duty and His Expectations* (London, 1749), vol. 1, pp. 341–51.

[7] For a discussion of LeSage's physics, see S. Aronson, 'The Gravitational Theory of George-Louis LeSage', *The Natural Philosopher* 3 (1964) 51–74; for a brief discussion of that theory's philosophical significance, see Chapter 6 below.

[8] From a letter published in *Notice de la Vie et des Ecrits de George-Louis LeSage* (ed. P. Prevost, Génève, 1805), p. 390.

[9] Ibid., p. 300.

[10] Ibid., p, 265.

[11] Ibid., p. 464–65.

[12] This quotation is from LeSage's 'Premier Mémoire sur la Méthode d'Hypothèse', published posthumously in P. Prevost's *Essais de Philosophie* (Paris, 1804), vol. 2, para. 23.

[13] See his *L'Art d'Observer* (2 vols. Geneve, 1775), expanded to the three-volume *Essai sur l'Art d'Observer et de Faire des Expériences* (Génève, 1802). Senebier, incidentally, was LeSage's immediate successor as Director of the Geneva Library.

[14] See P. Prevost, op. cit., note 12.

[15] D. Stewart, op. cit., note 5, vol. II, p. 301. (Cf. also ibid., pp. 307–308.)

[16] See Richard Olson's interesting study, *Scottish Philosophy and British Physics, 1750–1880* (Princeton, 1975).

[17] Cf. Mill's chapter on hypotheses in the *System of Logic.*

[18] See J. Mittelstrass, *Die Rettung der Phänomene* (Berlin, 1962).
[19] See M. Mandelbaum, *Philosophy, Science and Sense Perception* (Baltimore, 1964).
[20] See A. I. Sabra, *Theories of Light from Descartes to Newton* (London, 1967); and G. Buchdahl's numerous studies of Descartes.
[21] For a brief discussion of this issue, see Chapter 13 below.

CHAPTER 3

A REVISIONIST NOTE ON THE METHODOLOGICAL SIGNIFICANCE OF GALILEAN MECHANICS

It has long been common for scholars to maintain that the science of Galileo posed most of the central philosophical and methodological problems for early modern philosophy. Historians as diverse in orientation as Whewell and Mach, Koyré, and Cassirer have seen in Galilean physics the roots of the major philosophical problems of early modern science.[1] Doubtless, there are some very important philosophical issues raised by Galilean physics (among them: why nature is quantifiable, the role of ideal cases in physical theory, the character of thought experiments). But it is seriously misleading *historically* to assert that the philosophical issues implicit in Galileo's mechanics – or in *any* science like Galileo's mechanics – were the central source of intellectual *Angst* for scientifically-minded philosophers of the 17th and early 18th centuries. I shall suggest briefly and tentatively that we must look elsewhere in the scientific revolution to find those doctrines about nature which were responsible for the chief features of the 'epistemological revolution' of the early modern era.

If we want to determine the core epistemological and methodological problems of the new science, then we must look to the writings of the major philosopher-scientists of the period – to Descartes, to Hobbes, to Boyle, to Locke, to Newton, and to Leibniz. It was, after all, such thinkers as these who were primarily concerned to lay the new intellectual foundations for 17th-century science. Their central worries were *not* primarily worries about Galilean mechanics, nor yet worries about the astronomy of Copernicus which was, in turn, the *raison d'être* for Galileo's science of motion.

Rather, the central epistemological and methodological concern of most 17th-century philosophers was that of justifying certain scientific theories which were radically different from those of Galileo. What were these theories, and how did they pose different kinds of philosophical problems from those generated by Galilean mechanics? It is these questions to which I want to sketch an answer here.

It is crucial that we first remind ourselves that the sciences of astronomy and mechanics were not the only important sciences in the 17th century. Optics, chemistry, physiology, meteorology, and pneumatics (to name only a few) were also important sources of scientific research and theorizing. What is significant here is that *each* of these latter sciences raised epistemological and methodological issues which neither Galilean mechanics nor Copernican astonomy posed. Much of the philosophizing that went on in the 17th century addressed itself to such problems.

What is unique to astronomy and mechanics that left them out of mainstream philosophical inquiry? In brief, the nature of their subject matter and the sorts of theories appropriate to them. Both planetary astronomy and terrestrial mechanics are *macro-sciences*; they deal with properties and processes which can be more or less *directly* observed and measured. The apparent position of a planet, the rate of fall of a heavy body, the period of a pendulum – these are directly accessible phenomena. More to the point, the theoretical and explanatory machinery of Galilean mechanics deals chiefly with entities which are at least, in principle, observable. Galileo's various laws, which constitute the core of his explanatory repertoire, address themselves to observable entities and processes – or at least to ones that closely approximate in behavior those which we can observe. As Descartes pointed out frequently, Galileo generally refused to speculate seriously about many of the unseen (and presumably imperceptible) causal mechanisms which cause observable bodies to obey the laws which they do.

Alexandre Koyré has made much of Galileo's use of ideal cases (e.g., taking fall in vacuo as paradigmatic) and has argued that Galilean mechanics thereby posed a profound challenge to the empiricist epistemology championed by Aristotle and the Schoolmen. This seems to me to be a serious exageration, particularily when one contrasts Galilean mechanics with several of the other sciences of the 17th century. As Galileo himself goes to great length to show in many of his ingenious *Gedankenexperiments*, his ideal cases can often be approximated as closely as one likes in real life laboratory situations. If we take Aristotelian epistemology to be summed up in the dictum "nothing is in the mind which was not first in the senses", there is little in Galileo's science of motion which *need* be taken as challenging that epistemology. This is not to suggest, of course, that Galileo's own methodology was derivative from Aristotle's. Serious scholars continue to fight that one out. What is being claimed is that Galilean mechanics could be (and sometimes was) regarded as posing no acute threat to the theory of scientific methodology advocated (say) in Aristotle's *Posterior Analytics*. If the whole of 17th-century science had exhibited the largely phenomenological character of Galileo's mechanics, there need have been no revolution in methodology.

The picture is quite different, however, when we turn to the other sciences of the time. Although theories of optics, magnetism, capillarity, chemical change, and the like, address themselves to the explanation of observable phenomena, *the theories themselves postulated micro-entities which were regarded as unobservable in principle*. When Descartes, Hobbes, Hooke or Newton sought to explain, for instance, the refraction of light through a prism, their explanations invoked imperceptible particles of various sizes and shapes. When Boyle, Gassendi, or Descartes tried to explain chemical processes, they looked to 'the invisible realm' (as it was called by Newton) for their explanans. One of the most persistent, and philosophically disturbing, features of most sciences of the 17th century was the *radical observational inaccessibility of the entities postulated by their theories*. As numerous scholars have shown, it was the epistemological features of this type of theory which occasioned much of the philosophizing of the 17th and 18th centuries.

But where does Galileo fit into these concerns? The simple answer is that he does not. Although he himself certainly had views about the micro-structure of natural objects, the theories and principles for which he was best known in the 17th century manifestly did not raise epistemological problems of this kind.

Closely related to the problem of the unobservability of the entities postulated by most theories in the 17th century is another central problem. The theoretical entities postulated by Descartes, Boyle, Gassendi, Hobbes, Hooke, Huygens, and Newton had the following philosophically perplexing feature: they differed radically in their *properties* from macroscopic bodies. A ball of wax, to use Descartes' example, has a certain color, a certain smell, a certain texture, a disposition to melt, and the like. By contrast, the theoretical entities of the so-called corpuscular and mechanical philosophers exhibited *none* of these properties. Hence the philosophical conundrum: how can one determine which properties can be rationally attributed to these theoretical entities and which cannot? It is just this question which gave such prominence to the primary/secondary qualities distinction; a distinction which looms large in the philosophical writings of thinkers as different as Descartes and Newton, Boyle and Locke, Hobbes and Leibniz.

Here again, we must say that Galilean science – at least as it was perceived by his successors – did not raise this kind of philosophical worry. It is true that, in a now-famous passage in *The Starry Messenger*, Galileo made a distinction among the properties of bodies which is akin to the primary/secondary qualities distinction; but that distinction is largely gratuitous within the framework of Galilean mechanics or astronomy. His theories for those sciences neither entail nor presuppose that distinction.

Another way of putting all this is to say that the chief role of epistemology in the 17th century was that of *re-defining the character of the relation between theoretical knowledge and sensory experience*. Earlier epistemologists of science from Aristotle to Bacon had maintained that scientific theories could be elicited *from* nature by a careful and conscientious search for the "universals inherent in the particulars of sense". Precisely because Galilean mechanics could be (and often was) regarded as a natural extrapolation from sensory particulars, it posed few problems for the traditional epistemology of science. But the same could not be said for many of the other sciences of the 17th century. Unlike mechanics or astronomy, where one could imagine (in principle) that the basic postulates of the theory could be gleaned from experience, micro-physics and micro-chemistry left not even the remotest hope that one was doing anything but speculating *a priori*. There could, in short, be no plausibility at all in the claim that one could elicit one's first principles and initial conditions from nature itself. The mechanical philosophy was too palpably *postulational* and *hypothetical* for anyone to feign that pretense. As a result, the traditional mode of legitimating one's physical theories – that of claiming that they grew inexorably out of sensory experience – was foreclosed and scientists of an empiricist persuasion had to find other ways of getting experience (and experiment) into the knowledge system. If not at its beginnings, most of them reasoned, then at its end; if we cannot establish our first truths by deriving them *from* experience, the only seeming alternative is to test them (via their consequences) *against* experience. *Experience and experiment thus assume, for the mechanical philosophers, the role of testing our theoretical science, of authenticating its claim to be 'about' this world*. Such authentication was not needed in the classical image of science, precisely because it would have been redundant. If one has already gleaned one's first principles *from* nature, then any further comparison *with* nature is superogatory; for the first principles, if appropriately derived from sense, must be such that they correspond to nature. But for mid 17th-century scientists, that initial link with experience is lost and the only mode of theory authentication is to be found in an *a posteriori* comparison of theory with experience.

It might be objected at this point that there is really nothing very different about the methodological difficulties posed by Galilean mechanics and those posed by the corpuscular philosophy. In both cases, as we would now see it, one has a theory, freely created by the mind, which is then compared against nature. Logically, all the same problems occur. But to put it this way is to miss the central historical point that phenomenalistic systems of mechanics (such as those of Galileo) *were not seen to be incompatible* with the doctrine that our theories are elicited directly from nature. (Recall how much more difficult it was for Newton to argue for the experimental origins of his optical and chemical hypotheses than it was to argue that his mechanics had been "deduced from the phenomena".)

Let me try to highlight this point by putting it another way. I would want to argue (with suitable qualifications) that the central methodological shift associated with early modern science involved a recognition of the postulational and hypothetical character of theoretical knowledge. That recognition, in turn, was closely linked to a displacement of the experimental warrant for science away from what Reichenbach called the 'context of discovery' and towards the 'context of justification'.[2] The function of experience, on the earlier view, was to lead us to perceive the general (causal) connections between events and processes. Theory was seen simply as a generalized summary of what experience had already made clear; theorizing was a process of perceiving the universals in the midst of the particulars. Galilean mechanics posed no particular threat to this older view. Particular observations were, quite literally, *instantiations* of the phenomenological laws being tested. To make measurements on a ball rolling down an inclined plane was to get a form of direct accessability and direct testability of a sort which micro-explanatory theories in optics and chemistry could never lay claim to.

On the later view, however, the conception of experience as a generator of knowledge had been rejected; its role was no longer that of inspiring our theories, but that of providing *post hoc* checks on our freely-constructed hypotheses. This shift is of the *utmost* significance; it is tied up with the recognition of the fallibility of our knowledge, with the abandonment of a notion of productive causation, and with many other very important shifts in the history of the epistemology of science.

The micro-physical hypotheses of the 17th century were, I claim, one of the central sources for this shift from the view of experience as theory-generator to the view of experience as theory-tester. Phenomenalistic theories of whatever kind could not create the same embarrassments for the classical theory-generator view as micro-physical theories could. Proponents of the theory-generator view, confronted with the mechanics of a Galileo or a Newton, could (and did) still argue that its first principles had been gleaned from experience. (It was for just this reason that 18th-century followers of Bacon, who were as wedded to the theory-generator view as Aristotle had been, pointed with pride to the history of mechanics and with dismay to the history of micro-physical speculation.) No such argument could plausibly be constructed for atomic or aether theories, for their postulational and trans-empirical character was transparently clear. This interpretation is borne out by the fact that it was scientists working with micro-physics rather than those doing macro-physics who first began systematically to criticize the classical view that theories somehow arise out of the data. It is, after all, Descartes and Boyle (among others) who were really forced by the science they were doing to repudiate the earlier view. Although (as Mittelstrass has cogently argued) Galileo obviously used a postulational method, and was aware of doing so, it was not necessary to construe his mechanics in a postulational or hypothetico-deductive way. Galilean science, in all its dimensions, thus posed few of the threats to the classical methodology of theory construction which were posed by the micro-theoretical hypotheses of the later 17th and 18th centuries.

There is a more general issue at stake here, which goes well beyond the specific matter of Galileo. For almost three centuries, the historiography of the scientific revolution has been based on the assumption that astronomy and mechanics were the philosophically exciting sciences in the 17th century and that other sciences were somehow *parasitic* upon the Copernican-Galilean revolution in astronomy and mechanics. Recent scholarship, however, points to a different historiographical perspective. To wit, that it was not Copernicus's theory of the heavens, nor Galileo's theory of motion which produced the deepest transformations in the epistemology of the 17th century. Rather, the evidence strongly suggests that it was primarily *theories of matter* which posed the deepest threats to the traditional conceptions of the world and of knowledge, and which provided the real spark that kindled the 'philosophical revolution' of the 17th and 18th centuries. This, in turn, suggests that our historical focus, in coming to grips with the philosophical ramifications of the scientific revolution, should be directed less at figures like Copernicus and Galileo, and more at thinkers like Descartes, Gassendi, Boyle, and Hobbes; for it was the latter, and not the former, who posed the central epistemological and methodological problems of science for the next two centuries.

NOTES

[1] Cf. For a sampling of claims about the philosophical significance of Galilean science, see: A. Koyré, *Etudes galiléennes* (Paris, 1943); A. Koyré, *Metaphysics and Measurement* (London, 1968); E. Cassirer, *Das Erkenntnisproblem* (Berlin, 1911–20); E. A. Burtt, *Metaphysical Foundations of Modern Science* (London, 1932); M. Hesse, 'Galileo and the Conflict of Realism and Empiricism', in *Galileo nella storia e nella filosofia della scienza* (Piza, 1964); J. Mittelstrass, 'Two Concepts of Experience', forthcoming in

a volume to be published by Reidel; and W. Shea, *Galileo's Intellectual Revolution* (London, 1972).

It has been a common bias since the time of Kant that Galileo's science was the catalyst that produced early modern philosophy. It is a reasonable conjecture that the myth about Galileo's significance for the history of philosophy may owe much to the highly idiosyncratic character of Kant's conception of the philosophical problems of science.

[2] For a more detailed discussion of the linkages between a 'postulational' view of science and the logic of discovery, see Chapter 11.

CHAPTER 4

THE CLOCK METAPHOR AND HYPOTHESES: THE IMPACT OF DESCARTES ON ENGLISH METHODOLOGICAL THOUGHT, 1650–1670

THE CARTESIAN BACKGROUND

My tasks in this chapter are two-fold: to trace the influence of Descartes on 17th-century philosophy of science in Britain; and to document the fortunes of the method of hypothesis in Britain in the period immediately before Isaac Newton banished hypotheses from natural philosophy. These two tasks are not unrelated. Indeed, if what follows is anywhere near the mark, the two tasks collapse into one, for insofar as British thinkers of the period from 1650 to 1670 see any promise in the hypothetical method, it is because Descartes, by argument and example, enabled them to see it.

Historians have never been able to come to any very satisfactory conclusions about the influence of Descartes on 17th-century English thought. Until recently, it was thought that his impact was slight, significant – if at all – only in theology. In the last several decades, however, historians of science have detected Cartesian strains in English mechanics, optics, and physiology dating from the 1650s.[1] Gradually, therefore, the real and substantial role of Descartes is coming to be more fully appreciated. However, there is still one aspect of English philosophico-scientific thought where Descartes' positive impact is thought to be negligible, viz., with respect to theories of scientific method. Indeed, most historians who have dealt with the development of scientific method in Britain have written as if the 17th-century could be understood largely as a series of footnotes to, and commentaries on, Bacon's *Novum Organum*. Not only is 17th-century English philosophy of science said to be Baconian, it is equally thought to represent a violent reaction against the *a priori* Cartesian model of science, with its emphasis on all-embracing systems. These two factors, veneration for Bacon and scorn for Descartes, are allegedly the major stimuli for English writings on method from Hobbes to Newton. But this account has some profoundly disquieting features, not least of which is the fact that many English scientists and methodologists of this period were as vocal in their esteem for Descartes as in their idolatry for Bacon.

More significant, however, is the fact that several English natural philosophers of the period suggested that their accounts of scientific method were derived from, and perfectly compatible with, Descartes' views on the subject. Unless such thinkers were seriously misled about the origins of their ideas, we must critically re-examine the view of scholars such as R. F. Jones who insist that Descartes' methodological ideas had negligible impact compared with Bacon's.[2] There are, of course, well-established precedents for Jones' claim. The experimental tenor of the early Royal Society and its almost pathological aversion to hypothetical system-building seem to be symptoms of a latent, but well-entrenched, anti-Cartesianism. Furthermore, Thomas Sprat, in his influential *History of the Royal Society* (1667), extols the virtues of the experimental philosophy and barely mentions Descartes, except as an example of bad physics. This account is further reinforced by the lip-service which most British methodologists paid to Bacon, constantly speaking in exemplary tones of 'the noble Verulam', 'our illustrious Lord Bacon', etc. But despite such plausible precedents, this picture of Bacon as the sole or even dominant guiding light of British philosophy of science is too one-sided and seriously oversimplifies the diversity of the origins of English methodology in this period. While it is certainly true to say that English methodologists generally endorsed Baconian experimentalism,[3] it is not correct to think that they all accepted his inductive methods as well. Many thinkers were quite sceptical about the possibility of discovering indubitable scientific principles by any quasi-inductive process. In opposition to Bacon, they enthusiastically accepted Descartes' suggestion that the scientist must be content with hypothetical principles and conjectures rather than true and valid inductions. Descartes' hypotheticalism, blended with Baconian experimentalism, became a cornerstone of the methodologies of several English philosophers, especially Boyle, Glanvill, and Locke. In the general enthusiasm for Bacon, however, Descartes' contributions to English methodological thought have been neither documented nor carefully assessed. This chapter seeks to do just that.

I will argue, in contrast to Jones and others, that several of the major British natural philosophers of the mid-17th century derived their philosophies of science as much from Descartes as from Bacon. It follows, as a corollary to this, that they were neither so inductive nor so opposed to speculation as has often been suggested. I will claim that Descartes' methodology (especially that developed in the latter half of the *Principles*) was a fertile source for discussions of method among the English thinkers; and especially that his view of the universe as a 'mechanical engine' or clock whose internal parts can only be conjectured about served as an important stimulus for the English writers on method.

Before we can understand the debt of the English hypothetico-deductivists to Descartes, we must clarify the sense in which Descartes' methodology can be characterized as 'hypothetical'. Such emphasis has been placed on the *a prioristic* method, which he espouses in the *Discourse on Method*, that it may seem strange to suggest that he believed hypotheses to be indispensable to science.[4] But if we look to Descartes' scientific works or to his *Principles*, a very different picture emerges. More importantly, we must read the *Principles*, as the Englishmen of his time did, without bringing to it the prejudices that come from excessive pre-occupation with the *a priorism* of the *Meditations*. The figures I will be discussing knew Descartes primarily through the *Principles*[5] and it was thus natural for them to assume that he adopted a modest pose about the possibility of certainty in science, rather than the vain and omniscient posture of the *Discourse*.

Toward the end of the Fourth Part of the *Principles* (1644), Descartes makes a surprising confession. After trying to deduce the particular characteristics of chemical change from his first principles (i.e., matter and motion), he concedes failure. His program for the derivation of the phenomena of chemistry and physics from *a priori* truths remains uncompleted. His first principles are, he admits, simply too general to permit him to deduce statements from them about the specific way particular chunks of matter behave. He does not believe that bodies behave in violation of these first principles; Descartes was too confident, and his principles too vague, for him to be forced to admit that. But the very generality of his principles made them practically useless for explaining and predicting particular events.[6] Not content to leave anything unexplained, Descartes departed from his usual devotion to clear and distinct ideas and advocated the use of intermediate theories (less general than the first principles, but more general than the phenomena), which were sufficiently explicit to permit the explanation of individual events and which were, at the same time, *compatible with, but not deducible from, the first principles*. Descartes recognized that all such intermediary theories were inevitably hypothetical. Because their constituent elements were not clearly and distinctly perceived, it was conceivable that they were false. After all, nature is describable in a wide variety of ways and the fact that an explanation worked was no proof that it was true. Like any good logician, Descartes realized that "one may deduce some very true and certain conclusions from suppositions that are false or uncertain".[7]

Descartes goes on to suggest that we do not need assurances of truth in such matters. It will suffice if we can give an account of how nature might behave, not necessarily how nature does behave. After all, his was a corpuscular philosophy which sought to explain the macroscopic world in terms of imperceptible particles. Because such particles were unobservable, any specific properties attributed to them (e.g., such-and-such a size, shape, and motion) can only be so attributed tentatively and with a recognition of their hypothetical character. We can, of course, be sure that these micro-particles have *some* size, shape and motion (our first principles guarantee that much), but we remain forever in doubt about the particular properties they have.[8] Descartes justifies this excursion into the hypothetical by means of a metaphor which was widely exploited by later English writers who, as corpuscularians like Descartes, wanted some rationale for their use of hypotheses. He suggests that we imagine the world on the analogy of a watch, whose face is visible but whose internal construction is forever excluded from view. The most we can say about the inner workings of such a watch is conjectural opinion, not infallible knowledge. We can propose mechanisms concerning how the internal parts of the watch might be arranged, although we can never, *ex hypothesis*, get inside to see if we are right. Because the watch might be constructed in any number of ways, it is sufficient if we can outline some possible arrangement which would account for its external behaviour (e.g., hands moving, cuckoos calling and bells chiming). In the same way, the natural philosopher has honored his commitments so long as the mechanisms he proposes are compatible with the phenomena at hand. To ask for more than this is to misunderstand the limitations on natural inquiry. What Descartes is confronting here is a classic species of *empirical underdetermination* of theories. He believes that indefinitely many mutually inconsistent micro-structural hypotheses are all compatible with the visible effects. Under such circumstances, one cannot conceivably hope to use experience to pick out uniquely the correct hypothesis. The empirical constraints from below and the ontological demands from above still leave, in Descartes' view, a large range of choice between rival hypotheses – all of which satisfy all the available constraints. Descartes illustrates this epistemic conundrum with a lengthy analogy. The crucial part of the passage reads as follows:

> It may be retorted to this that, although I may have imagined causes capable of producing effects similar to those we see, we should not conclude for that reason that those we see are produced by these causes; for just as an industrious watch-maker may make two watches which keep time equally well and without any difference in their external appearance, yet without any similarity in the composition of their wheels, so it is certain that God works in an infinity of diverse ways [each of which enables Him to make everything appear in the world as it does, without making it possible for the human mind to know which of all these ways He has decided to use]. And I believe I shall have done

enough if the causes that I have listed are such that the effects they may produce are similar to those we see in the world, without being informed whether there are other ways in which they are produced.[9]

The clock analogy is no afterthought which Descartes threw in to illustrate his underdetermination argument. Rather, it formed an integral part of his way of looking at the world and the role he assigned to the corpuscular philosophy in explaining that world. He tells us that machines like the clock served as models for developing his mechanical account of nature:

And in this, the example of certain things made by human art was of no little assistance to me; for I recognize no difference between these machines and natural bodies . . . .[10]

To understand the significance of the clock analogy and why it led Descartes to advocate the method of hypothesis, we must look carefully at his account of scientific knowledge. Though he frequently speaks of deducing the facts of physics from his first principles,[11] he never offers any deduction which does in fact exhaustively or uniquely explain some particular in terms of these very general principles. We need many additional assumptions beyond the 'first principles' to explain why observable bodies behave the way they do, and these assumptions cannot all be derived from the first principles. Indeed, whenever Descartes actually tries to deduce optical and mechanical phenomena from the first principles, he persistently fails and must fall back on a variety of hypothetical assumptions. Nor should we be startled to find that the matter-in-motion paradigm is too austere to enable us to explain particular events. After all, every body has matter and a state of motion, but only some are luminous, magnetic, dense or abrasive. Clearly something else is involved which gives matter these characteristics. Wanting to avoid any occult forces (because only matter and motion truly exist), Descartes finds that the only way to explain such properties as light and magnetism is to assume that bodies exhibiting these properties have a different sort of motion from bodies which do not have these properties.[12] Despite Descartes' claim that he could deduce everything in his optics from his clear and distinct ideas, he is continually forced to employ assumptions that do not follow from any knowledge we have of the first principles. He is compelled to make additional assumptions about the constitution of moving matter. Such assumptions form an essential part of every scientific explanation. Thus in, addition to the first principles, we need a set of principles of lower generality which will enable us to characterize the specific mechanisms of nature. What shapes, sizes, and motions are exhibited by the unobservable constituents of macroscopic bodies? Answers to such questions cannot be *derived* from the first principles.

More to the point, Descartes realized that these principles of intermediate generality were not deducible from his metaphysical strictures about nature. In this regard, Buchdahl has correctly noted that: "It is a scholar's legend that Descartes consistently believed that his physics was deducible from first principles".[13]

Descartes' endorsement of the hypothetical method is most explicit in that section of the *Principles* where he develops the doctrine of the three elements, which he used extensively to explain chemical and physical change. Among Descartes' assumptions is the claim that matter is corpuscular and that these corpuscles have a certain size, shape, and velocity. He says of these assumptions:

> we cannot determine by reason how big these pieces of matter are, how quickly they move, or what circles they describe . . . [this] is a thing we must learn from observation. Therefore, we are free to make any assumptions we like about them, so long as all the consequences agree with experience . . . .[14]

Here again, matter and motion are too general to explain the phenomena. We must, he insists, resort to less general hypotheses about the size and configuration of matter in order to explain the world. The scientist thus resembles the skilled watch-maker of the analogy who is given a watch but cannot see its internal mechanisms. Like the watch-maker, he knows the general principles which govern his subject matter, but he is uncertain about the way they exhibit themselves in any particular case. Equally like the watch-maker, the scientist can offer only conjectures about internal construction and mechanisms.

The role of the first principles in physics is thus to *circumscribe* the range of acceptable hypotheses by excluding certain entities. Our first principles tell us, for example, not to develop a science based on the assumption of a void; they warn us against hypotheses couched in the teleological language of final causality; and they forbid hypotheses postulating action at a distance. Viewed in this light, corpuscular metaphysics does not dictate which physics we adopt, but only gives us certain regulative constraints. Physical hypotheses must be compatible with such regulative principles, but they are manifestly not deducible from them.

But though Descartes concedes that science is necessarily hypothetical and probabilistic, he is not willing to say that all hypotheses are equally good or that the scientist can never be confident about his principles. He does suggest that mere *ad hoc* hypotheses, invented to explain one particular phenomenon, are not very convincing. But, he insists, when we put forward a hypothesis which accounts for a wide variety of phenomena successfully, we can be reasonably confident (though not certain) that it is true:

> Although there exist several individual effects to which it is easy to adjust diverse causes [i.e., hypotheses], one to each, it is however not so easy to adjust one and the same [hypothesis] to several different effects, unless it be the true one from which they proceed.[15]

While insisting on the conjectural character of scientific hypotheses, Descartes was careful not to succumb to the sceptic's temptation to grant all hypotheses equal status and improbability. He declared the right of the scientist to accept those hypotheses which accounted for a wide cross-section of the facts at hand because "it is not likely that that from which one may deduce all the phenomena is false".[16] Thus, a sound hypothesis is one which is compatible with the data and with the first principles, matter and motion. On Descartes' view, the logical gap separating the first principles from the phenomena can be bridged only by hypotheses. Since *compatibility* rather than deducibility is the relation between the first principles and the hypotheses of physics; the first principles function in the same way, *vis-à-vis* the hypotheses, as the facts do. Our first principles, like the data, can inform us that certain hypotheses are wrong; but they cannot tell us which hypotheses are right. We can never get inside nature's clock to see if the mechanisms are what we think them to be. However, the clock analogy is important not only for the considerable light it throws on Descartes' use of the method of hypothesis, but equally because it, or variants on it, were widely cited by subsequent writers who, as corpuscularians, were struggling with the same methodological problems.[17] In particular, it was used by many English writers (Boyle, Glanvill, Power, and Locke) who are generally regarded as Baconian experimentalists, uninfluenced by Descartes. The metaphor thus provides a convenient motif in terms of which to explain the development of the method of hypothesis between Descartes and Newton.

BOYLE AND CARTESIAN PROBABILISM

We now turn to consider the more general theme of this chapter, namely the impact of Descartes' method of hypothesis on English writers of the following generation.[18] His influence can be seen most prominently in the work of Robert Boyle, who did much to fuse the Baconian and Cartesian traditions into a coherent and sophisticated view of scientific method. Because Boyle's philosophy of science unifies major elements from both, it must be understood in the context of the traditions which those two writers initiated. In the mid-17th century, Baconianism and Cartesianism signified quite different things from what they do today. Bacon was not praised (or condemned) as an inductive philosopher so much as an experimental one. Descartes, on the other hand, was not treated as an *a priorist* but rather as an advocate of the corpuscular philosophy who encouraged the use of hypotheses in science. So far as one can judge from his published works, it never seriously occurred to Boyle that the principles of science could be discovered either inductively or *a priori*. By neglecting Bacon's inductivism and Descartes' rationalism, Boyle viewed the methodologies of these two authors as healthy contrasts of emphasis within a commonly-held scientific world-view.

Nor should we be startled to find resemblances between the 'experimentalist' Boyle and the 'rationalist' Descartes. After all and above all, they were both corpuscularians. Boyle began his philosophical education with heavy doses from Descartes;[19] and as we have seen, the Descartes of the *Principles* was neither so *a priori* nor so anti-experimental as we might now judge him to be on the basis of his *Discourse* and *Meditations*. The Descartes of the *Principles* is not the proponent of systematic doubt so much as the modest inquirer after truth who admits, expecially throughout the latter half of the *Principles*, that science is an hypothetical and conjectural enterprise which offers its followers only a probable story, not the revealed truth. Nor is it inconsistent for Boyle to draw from both Descartes and Bacon, for their methodologies, as Boyle construed them, were not contradictory. Indeed, Descartes' hypothetical method can be viewed (and was so viewed by Boyle) as an alternative formulation of Bacon's hypothetical Indulgence of the Understanding.[20]

Any attempt to explain Boyle's methodology must, of course, begin with the fact that he was an ardent adherent of the corpuscular or mechanical philosophy.[21] Indeed, it is his corpuscularism which conditioned his whole approach to nature and which inclined him to adopt Descartes' method of hypothesis, while simultaneously taking his experimentalism from Bacon. As an advocate of the corpuscular philosophy, and as a writer squarely in the experimental tradition, Boyle was acutely aware of the immense gap separating the principles of corpuscularism from particular sciences such as chemistry and physiology. Boyle perceived that this was a serious weakness in the mechanical program:

But I am sorry to see cause to add to what I have been saying, that as much as we [Corpuscular philosophers] magnify the necessity of experiments in our contacts with the Peripateticks about nature, we seem not yet to be sensible of this acknowledged necessity, when we contest with the particular difficulties that frequently occur, when we ourselves are to discover the cause of her phenomena, or to imploy her productions.[22]

If the corpuscular philosophy is to be useful to natural philosophy, it must do more than propose a few vague principles which it asserts to be compatible with nature. It must use those principles and others to explain what can be observed in the laboratory. It is not enough to say that fire boils water because the rapidly moving corpuscles of fire break up water clusters and send vapor to the surface. We must pay careful attention to describing the particular shape and velocity of fire corpuscles, to the mechanisms whereby they break up liquid clusters and to the laws of boiling. More generally, the corpuscular philosophy must cease to be a set of ambiguous metaphysical principles which are so fluid that they are compatible with any phenomenon. Boyle hopes to transform the corpuscular doctrine into a sensible physical theory which makes predictions and provides explanations; in short, into a theory which approaches experience in order to learn from it and which stakes its fate, not on the philosopher's ability to weave intricate myths and devise *ad hoc* adjustments, but on the scientist's ability to confirm those principles.

Boyle never seriously doubted that nature is ultimately matter-in-motion,[23] but he insists that we need to go beyond such cryptic formulae if we are to have a science worthy of the name:

For it is one thing to be able to shew it possible for such and such effects to proceed from the various magnitudes, shapes, motions, and concretions of atoms, and another to be able to declare what precise and determinate figures, sizes, and motions of atoms, will suffice to make out the proposed phenomena.[24]

We must formulate intermediary theories which are at once less general and more explicit than matter and motion. Like Descartes in the *Principles*, Boyle realized that the typical corpuscularian doctrine is too general to permit one to explain, in detail, the behavior of matter. We must develop lower-level theories which, while compatible with the corpuscular philosophy, are not strictly deducible from it:

There are a great many things which ... cannot with any convenience be immediately deduced from the first and simplest principles; namely, matter and motion; but must be derived from subordinate principles; such as gravity, fermentation, springiness, magnetism, etc.[25]

In the ideal case, we should seek to derive everything from the first principles.

Unfortunately, however, there is a wide discrepancy between what we hope for and what we are prepared to accept:

That we may aspire to, but must not always require or expect, such a knowledge of things, as is immediately derived from their first principles.[26]

Even where our explanations are not derivable from mechanical principles, they must be compatible with them:

the mechanical principles are so universal, and therefore applicable to so many things, that they are rather fitted to include, than necessitated to exclude, any other [subordinate] hypothesis, that is founded in nature, as far as it is so. And such hypotheses . . . will be found, as far as they have truth in them, to be legitimately (though perhaps not immediately) deducible from the mechanical principles, or fairly reconcileable to them . . .[27]

Every subordinate hypothesis, in so far as it is legitimate, is either deducible from, or at least compatible with, the corpuscular philosophy. Descartes made precisely the same point in the *Principles*. So far as natural philosophy is concerned, these less general hypotheses are even more useful than the concepts of matter and motion:

The most useful notions we have in physicks . . . are not derived immediately from the first principles; but from intermediate theories, notions, and rules.[28]

Boyle set himself the life-long task of enunciating such "sub-ordinate principles" and "intermediate theories"[29] in order to provide scientific flesh for the metaphysical skeleton of corpuscularism:

I thought it would be no slight service, not only to the Corpuscular hypothesis, but to natural philosophy itself, if I could by good experiments, and at least probable reasons, make out that almost all sorts of particular qualities may be mechanically originated or produced.[30]

Having settled on such an undertaking, it was natural that Boyle should give some thought to the method whereby these subordinate principles could be discovered and confirmed. It is at this stage that we see a remarkable blend of Baconian and Cartesian elements. With Bacon, Boyle emphasized that the proper foundation of physical knowledge was experimentation; not merely casual observation of nature, but systematic and often artificial tinkering with the physical world so as to observe it under a wide variety of circumstances. Good natural philosophers, he writes, consult experience both more frequently and more heedfully [than the Aristotelians]; and, not content with the phenomena that nature spontaneously affords them, they are solicitous, when they find it needful, to enlarge their experience by trials purposely devised . . .[31]

Like Bacon, he envisaged the compilation of vast histories of nature which would summarize and codify the information gleaned from experiment.[32] To this end, Boyle himself wrote experimental histories of fluidity, firmness, colors, cold, air, respiration, condensation, flames, human blood, porosity, liquors, tin, and fire. But what are we to do with such natural histories once they are compiled? Can we use them to induce, Baconian fashion, the principles and laws of science? Boyle's answer to this question is an unequivocal "No". Though a self-styled pupil of Bacon, Boyle never, to my knowledge, uses the term 'induction'[33] nor does he ever seriously consider Bacon's view that principles will emerge in any mechanical way from a study of nature.[34] So far is Boyle from Bacon's conception of a rigidly experimental science that he even suggests that a good hypothesis is more valuable than a well-conceived experiment:

And tho' perhaps few have a greater love and value for experiments than I, yet for my part, I should think myself more obliged to him that discovers to me some pregnant notion . . . than if he imparted some fine experiment.[35]

Far from following Bacon's inductive line, Boyle takes a more Cartesian pose. The purpose of all this experimentation, he tells us, is to place us in a position to offer some hypothesis as a tentative account of the data which we have so carefully accumulated.

The experimental histories thus serve as the raw material for the theorist who then proceeds to make conjectures which are tested in terms of their ability to "render an intelligible account of the causes of the effects, or phenomena proposed [in the histories]".[36] Boyle does not believe that theories will arise ready-made from the data, or that the data will uniquely determine any single theory.[37] The data are important because without them we might accept a theory which would have been falsified if we had experimented more thoroughly. But it is the faculty of reason which constructs theories from the data; they do not spring full-blow from the histories. No matter how extensive our experimentation, science remains fundamentally hypothetical.[38]

These ideas are similar, in expression as well as content, to Descartes' methodological position in the *Principles* and speak forcefully against those who minimize his impact on English thought in this period. But it might be argued that all I have said to this point merely suggests that Boyle and Descartes were both sympathetic to the method of hypothesis, but not necessarily that Boyle's version of that method was indebted to Descartes. This might seem just another of those coincidences that continually tantalize historians of ideas. That Boyle had similar views twenty years after Descartes is, of itself, meagre proof of his debt to Descartes. In history as well as logic, *post hoc* is no guarantee of *propter hoc*. Fortunately, however, this is not the only evidence we can cite for the claim that Descartes exerted substantial influence on English methodological thought, especially Boyle's. There is a passage in Boyle's *The Usefulness of Natural Philosophy* (1663) which makes the case a good deal more cogent. For Boyle there takes Descartes' clock analogy and, by clumsily paraphrasing it, uses it to justify – much as Descartes did – an avowedly hypothetical and corpuscular methodology. Boyle formulates the analogy thus:

> . . . many Atomists and other Naturalists, presume to know the true and genuine causes of the things they attempt to explicate; yet very often the utmost they can attain to, in their explications, is, that the explicated phenomena may be produced after such a manner, as they deliver, but not that they really are so. For as an artificer can set all the wheels of a clock a going, as well with springs as with weights,. . . so the same effects may be produced by divers causes different from one another; and it will oftentimes be very difficult, if not impossible for our dim reason to discern surely, which of those several ways, whereby it is possible for nature to produce the same phenomena, she [nature] has really made use of to exhibit them.[39]

Descartes was not the first to liken nature to a clock-like mechanism; on the contrary, this was a common metaphor among mechanistic philosophers throughout the 16th and early-17th centuries. The fact that English writers like Boyle also used the clock analogy is, of itself, no indication of their Cartesian leanings. However, Descartes was (so far as I can tell) the first to use the analogy to justify a hypothetical view of knowledge and science. That Boyle and others used the clock analogy *in precisely the same way Descartes did* – to buttress up a hypothetico-deductive methodology – is probably indicative of their Cartesian leanings. This clock metaphor, borrowed from Descartes, particularly struck Boyle's fancy; so much so that he remained persuaded for the rest of his life that science could attain only probable, not infallible knowledge. The language in this passage is Boyle's; but the thought is clearly Descartes'. Both insist that our theories only describe the mechanisms whereby nature might conceivably produce the effects we observe, not necessarily the mechanisms which nature in fact uses. In another passage reminiscent of Descartes, Boyle puts the point this way:

> ... it is a very easy mistake to conclude, that because an effect may be produced by such determinate causes, it must be so [produced], or actually is so.[40]

He shared with Descartes the belief that matter and motion are the ultimate and true principles of physical science, and he insisted, again like Descartes, that all subordinate principles, in terms of which we explain particular events, are necessarily conjectural. While declaring himself a faithful atomist, Boyle was very sceptical about the possibility of the 'Atomical Hypothesis' ever becoming more than a probable theory; and, for that matter, he was equally pessimistic about ever discovering any of the true mechanisms of nature.[41] Like Descartes, Boyle is careful not to confuse verification with proof. If a hypothesis has accounted for all the phenomena, then it has demonstrated its *utility* but its *truth* is still an open question and forever remains so. We may, by chance, stumble on to a true hypothesis, but we can never prove it to be so. He likens science to a deciphering operation in which:

> men conjecturally frame several keys [i.e., hypotheses] to enable us to understand a letter written in ciphers [i.e., nature]. For although one may by his sagacity have found the right key, it will be very difficult for him, to prove [it is the right one].[42]

Apart from the evidence based on similarity of language and theme, there are other factors which suggest that Boyle drew his hypothetical method from Descartes. Just as Descartes justified his hypothetical method by attributing it to Aristotle's *Meteors*,[43] so does Boyle turn to the Stagirite to show that the method of hypothesis had been anticipated by 'the master of those that know':

> Aristotle himself (whatever confidence he sometimes seems to express) does in his first book of *Meteors* ingenuously confess, that concerning many of nature's phenomena, he thinks it sufficient, that they may be so performed as he explicates them.[44]

Boyle's method then consists in this: The scientist conducts wide-scale experimentation to determine the "divers effects of nature". He next suggests a hypothesis to explain what has been observed. The first hypothesis should be a fairly low-level generalization about the "immediate causes of the phenomena". Then, "ascending in the scale of causes", he arrives ultimately at the most general hypotheses, which concern the "more catholick and primary causes of things".[45] At each level, the scientist checks to see if the hypothesis conforms to the corpuscularian doctrine. If so, he tests the hypothesis against the entries in all his tables and against the other known laws of nature. If it is falsified he rejects it; if not, he continues to maintain it. This is obviously similar to Descartes' view that the intermediate hypotheses must be compatible with the first principles and with the phenomena. An hypothesis is not proven true, of course, even if it is compatible with all our evidence; but it can be asserted with more confidence as it proves itself capable of explaining more and more phenomena:

For, the use of an hypothesis being to render an intelligible account of the causes of the effects, or phenomena proposed, without crossing the laws of nature, or other phenomena; the more numerous, and the more various the particles are, whereof some are explicable by the assigned hypothesis, and some are agreeable to it, or, at least, not dissonant from it, the more valuable is the hypothesis, and the more likely to be true. For it is much more difficult to find an hypothesis, that is not true, which will suit with many phenomena, especially if they be of various kinds, than but with a few.[46]

Here again, comparisons with Descartes are in order. Descartes had argued that those principles are most likely which explain 'several different effects' rather than one, and had suggested that a well-confirmed hypothesis, which explains a variety of instances, is probably true.[47] Boyle's point is substantially the same.

There is yet another basic methodological postulate which Boyle and Descartes both accepted. This might be called the principle of the *multi-level identity of nature*. Basically, this principle stipulates that the laws of nature which apply to visible bodies also apply to objects which are either too large or too small to be measured or observed.[48] It was by invoking such a principle that 17th-century scientists were able to assume that the laws of visible-body mechanics applied to interactions between sub-microscopic corpuscles. It was also in terms of this principle that they rejected the scholastic strategem of attributing properties to micro-entities which do not describe observable entities. Descartes formulates the principle thus:

we do much better to judge of what takes place in small bodies, which their minuteness alone prevents us from perceiving, by what we see occurring in those that we do perceive, than, in order to explain certain given things, to invent all sorts of novelties, that have no relation to those that we perceive.[49]

Boyle, adopting different language, puts the principle this way:

both the mechanical affections of matter are to be found, and the laws of motion take place, not only in the great masses, and the middle sized lumps, but in the smallest fragments of matter . . . And therefore to say, that though in natural bodies, whose bulk is manifest and their structure visible, the mechanical principles may be usefully admitted, [but] that [they] are not to be extended to such portions of matter, whose parts and texture are invisible; may perhaps look to some, as if a man should allow, that the laws of mechanism may take place in a town clock, but not in a pocket watch.[50]

By way of recapitulation of the argument thus far, I shall enumerate below the most important passages illustrating methodological similarities between Descartes and Boyle:

DESCARTES	BOYLE
1. 'And lest it be supposed that Aristotle did, or wanted to do, more than this, it must be recalled that he expressly says in the first book of the *Meteors*, at the beginning of the seventh chapter, that with regard to things not evident to the senses, he thinks that he offers sufficient explications and demonstrations of them, if he merely shows that they may be as he explains them.'	1. 'Aristotle himself (whatever confidence he sometimes seems to express) does in his first book of *Meteors* ingenuously confess, that concerning many of nature's phenomena, he thinks it sufficient that they may be so performed as he explicates them.'
2. 'For just as an industrious watch-maker may make two watches which keep time equally well and without any difference in their external appearance, yet without any similarity in the composition of their wheels, so it is certain that God works in an infinity of diverse ways . . . And I believe I shall have done enough if the causes that I have listed are such that the effects they may produce are similar to those we see in the world, without being informed whether there are other ways in which they are produced.'	2. 'For as an artificier can set all the wheels of a clock a going, as well with springs as with weights, . . . so the same effects may be produced by divers causes different from one another; and it will oftentimes be very difficult, if not impossible for our dim reasons to discern surely, which of those several ways, whereby it is possible for nature to produce the same phenomena, she has really made use of to exhibit them.'
3. 'Although there exist several individual effects to which it is easy to adjust diverse causes [i.e. hypotheses], one to each, it is however not so easy to adjust one and the same [hypothesis] to several different effects unless it be the true one from which they proceed.'	3. 'For it is much more difficult, to find an hypothesis, that is not true, which will suit many phenomena, especially if they be of various kinds, than but with a few.'
4. 'One may deduce some very true and certain conclusions from suppositions that are false or uncertain.'	4. 'It is a very easy mistake to conclude that because an effect may be produced by such determinate causes, it must be so, or actually is so.'
5. 'We do much better to judge of what takes place in small bodies, which their minuteness alone prevents us from perceiving, by what we see occurring in those that we do perceive, than, in order to	5. 'And therefore to say, that though in natural bodies, whose bulk is manifest and their structure visible, the mechanical principles may be usefully admitted, [but] that [they] are not to be extended to

explain certain given things, to invent all sorts of novelties, that have no relation to those that we perceive.'

such portions of matter, whose parts and texture are invisible; may perhaps look to some, as if a man should allow, that the laws of mechanism may take place in a town clock, but not in a pocket watch.'

The essence of Boyle's hypothetical method is falsification.[51] Like both Bacon and Hooke, Boyle insists on the importance of refuting hypotheses rather than confirming them. Like them, he compares scientific inquiry to the *reductio ad absurdum* methods of the mathematicians. Scientists, he writes, should try:

diligently and industriously to make experiments and collect observations, without being over-forward to establish principles and axioms, believing it uneasy to erect such theories, as are capable to explicate all the phenomena of nature, before they have been able to take notice of the tenth part of those phenomena, that are to be explicated. Not that I at all disallow the use of reasoning upon experiment . . . for such an absolute suspension of the exercise of reasoning were exceedingly troublesome, if not impossible. And, as in that rule of arithmetic, which is commonly called *regula falsi* [i.e., false position], by proceeding upon a conjecturally-supposed number, as if it were that, which we inquire after, we are wont to come to the knowledge of the true number sought for; so in physiology [i.e., natural philosophy] it is sometimes conducive to the discovery of truth, to permit the understanding to make an hypothesis, in order to the explication of this or that particular difficulty, that by examining how far the phenomena are, or are not, capable of being solved by that hypothesis, the understanding may, even by its own errors, be instructed. For it has been truly observed by a great philosopher [Bacon], that the truth does more easily emerge out of error than confusion.[52]

Science, as he conceives it, is a dialectic between reason and sensation; the mind proposes theories which are then subjected to the careful scrutiny of the senses. Boyle does not share Bacon's contempt for intellectual activities nor his distrust of 'philosophical systems', though he is quick to point out that:

such kind of superstructures [should be] looked upon only as temporary ones; which . . . are not entirely to be acquiesced in, as absolutely perfect, or uncapable of improving alterations.[53]

But so long as they are preceded by experimentation and never enunciated with finality, theoretical systems and hypotheses play a central role in science. Boyle is the first to admit that many mechanistic hypotheses appear, on first glance, unseemly and bizarre. But he insists that a deeper analysis will reveal their inherent value:

Some hypotheses may be compared to those shops of drugsters and apothecaries – where among the first things that the eye discovers, are serpents and crocodiles, and other monstrous and harmful things, whereas the inside is a repository or magazine of wholsome and useful medicines.[54]

Boyle's attitude towards the method of hypothesis comes out most clearly in his polemic against Hobbes' scathing critique of the Royal Society. Hobbes, like Boyle, of course, believed hypotheses to be indispensable to natural philosophy and he had little sympathy for those in the Royal Society who were so obsessed with the new experimental philosophy that they denied any role to speculations or conjectures. Boyle replies to Hobbes' critique by arguing (in a vein not quite faithful to the trenchant empiricism of many members of the Royal Society) that the new organization was not opposed to the use of all hypotheses, but only to those that are not well grounded in the phenomena. Hobbes had proposed two requisites of a good hypothesis: that it be conceivable and that, if it is granted, "the necessity of the phenomena may be inferred".[55] To these, Boyle adds a third and equally important condition: "namely that it [the hypothesis] not be inconsistent with any other true or phenomenon of nature".[56] Elsewhere, he formulates the same condition another way: it is sufficient to accept an hypothesis if we can demonstrate "its fitness to solve the phenomena for which [it was] devised, without crossing any known observation or law of nature".[57] Richard Westfall[58] and Marie Boas Hall[59] have pointed out that in Boyle's unpublished manuscripts, there is an even more explicit discussion of the characteristics of an acceptable hypothesis. Boyle writes that the requisites of a 'good hypothesis' are (1) that it suppose nothing that is either impossible or absurd, (2) that it be self-consistent, (3) that it be sufficient to explicate the phenomena, "at least the chief among them", and (4) that it be consistent with other known phenomena and "manifest physical truth". An "excellent hypothesis", in addition to satisfying these four requirements, also (5) is the simplest hypothesis, (6) is the only hypothesis that explicates the phenomena or at least explicates them better than any other and (7) enables us to "foretell future phenomena".[60]

What I have said here should not be taken to suggest that Boyle's methodological ideas are identical with Descartes'. There are important differences between Descartes and Boyle which should not be glossed over. Boyle's continual insistence on extensive experimentation finds no counterpart in Descartes and he had no patience with Descartes' exclusion of final causes from cosmology. Furthermore, Boyle often chastises Descartes, not for constructing hypothetical systems, but for constructing them with insufficient care. He also notes, rightly, that Descartes paid only lip-service to experience and spun his corpuscular philosophy largely out of his head. But just as the differences between the two should not be neglected, neither should their similarities, especially when they are so pronounced as they are in regard to the method of hypothesis.[61] In the methodological writings of Boyle, Descartes' influence is every bit as pronounced as Bacon's.

Boyle was not only aware, as most of his compatriots were, that the Cartesian system was composed of hypotheses. He also appreciated, as too few of his countrymen did, Descartes' *reasons* for using hypotheses. In short, he understood Descartes' claim that hypotheses were essential to science, that they were the sign of a fertile mind aware of nature's inscrutability. Although Boyle often disagreed with the specific hypotheses which Descartes formulated, he never made the Newtonian mistake of thinking that the Cartesian method of hypothesis must be rejected because certain Cartesian hypotheses were ill-founded.

Nor should we be surprised that Boyle and Descartes were of one mind concerning the method of hypothesis. Once once accepts the corpuscularian theory of matter, and with it the theory of knowledge which makes corpuscles unobservable in principle, then it is altogether natural to espouse a version of the method of hypothesis. In short, the metaphysics and epistemology of the mechanical philosophy led, by its own inner logic, to the advocacy of a certain methodology.[62] Boyle accepted the Cartesian theory of method, not because he was a Cartesian, but because he was a corpuscularian!

In a sense, Boyle borrowed the best from both worlds. Impressed with the clock analogy, he sensibly appropriated Descartes' method of hypothesis, but neatly dropped the doctrine of clear and distinct ideas which regularly plagued the development of Descartes' views on method. From Bacon, he inherited a vigorous empiricism, but was careful not to import with it the inductive and rigidly 'empiricist' philosophy which cramped Bacon's acceptance of the hypothetical method and made his endorsement of it invariably half-hearted.[63]

GLANVILL AND CARTESIAN PROBABILISM

A second English writer much impressed by Descartes' clock-metaphor was Joseph Glanvill (1638–1680) who, in his *Vanity of Dogmatizing* (1661) and *Scepsis Scientifica* (1665), has some illuminating, if epistemically depressing, things to say about scientific method.[64] Though both volumes are declarations of war against classical and medieval philosophy, Glanvill has kind words for at least two of his predecessors, Bacon and Descartes. Like Boyle, he drew from the two authors as if they were merely different sides of the same coin. He frequently alludes to "those great men, the Lord Bacon and Descartes"[65] and his debt to each of them will become clear as we describe his methodology.

Nature, writes Glanvill, is very subtle and its mechanisms are not obvious to even the most astute observer. Only vain and pompous men believe that their hypothetical scientific systems provide a faithful image of the physical world. Natural philosophy is seriously mistaken when it dictates how the physical world should operate. To deny a void or the earth's rotation because we conceive such things to be impossible is to make the error of thinking that nature and its Creator are confined within the bounds of man's dim reason; which is not only absurd, but blasphemous, to the pious Glanvill. At this point, Glanvill invokes Descartes' analogy:

For Nature is set a going by the most subtil and hidden instruments; which it may have nothing obvious which resembles them. Hence judging by visible appearances, we are discouraged by supposed impossibilities which to nature are none, but within her spear [sic] of action. And therefore what shews only the outside, and sensible structure of nature; is not likely to help us in finding out the *Magnalia* [i.e., inner mechanisms]. 'Twere next to impossible for one, who never saw the inward wheels and motions, to make a watch upon the bare view of the circle of hours, and index: And 'tis as difficult to trace natural operations to any practical advantage, by the sight of the cortex of sensible appearances.[66]

Sensation alone can never give us insight into the true mechanisms of nature; the most we can realistically hope for is probable and hypothetical knowledge. Thus, the "way of inquiry for true philosophers" is

to seek truth in the great book of nature, and in that search to proceed with wariness and circumspection without too much forwardness in establishing maxims and positive doctrines: To propose their opinions as hypotheses, that may probably be true accounts, without peremptorily affirming that they are.[67]

Glanvill leaves no doubt that he borrowed his hypotheticalism from Descartes. Towards the end of the *Scepsis Scientifica*, he writes:

And though the Grand Secretary of Nature, the miraculous Descartes hath here infinitely out-done all the philosophers that went before him . . . yet he intends his principles but for hypotheses, and never pretends that things are really or necessarily, as he hath supposed them: but he only claims that they may be admitted pertinently to solve the phenomena, and are convenient supposals for the use of life. Nor can any further account be expected from humanity, but how things possibly *may have been made* consonantly to sensible nature: but infallibly to determine how *they truly were effected*, is proper to him only that saw them in the chaos, and fashioned them out of that confused mass. For to say, the principles of nature needs be such as our philosophy makes them, is to set bounds to omnipotence, and to confine infinite power and wisdom to our shallow models.[68]

This statement alone is sufficient to undermine the claim that Descartes' methodology had no impact on the English writers of this generation. It makes clear that Descartes' probabilism, and the clock analogy which supports it, were obviously attractive to a significant group of Englishmen who, for religious or other reasons, saw fit to put strict limitations on the scientist's access to truth.[69] Glanvill, however, carries his probabilism further than Descartes, he denies that *any* scientific principles can be more than hypotheses.

For the best principles, excepting divine, and mathematical, are but hypotheses; within which, we may conclude many things with security from error. But yet the greatest certainty, advanced from supposal, is still but hypothetical. So that we may affirm, that things are thus and thus, according to the principles we have espoused: But we strangely forget ourselves, when we plead a necessity of their being so in nature, and an impossibility of their being otherwise.[70]

Glanvill then is an uncompromising advocate of the method of hypothesis, more consistently so than either Descartes or Boyle. But while insisting that all scientific principles are but conjectures, he believed that some hypotheses are better than others. Taking Boyle's view, Glanvill argued that sound hypotheses must be based on experimental histories of nature.[71] To the Royal Society, he wrote:

... from your promising and generous endeavors, we may hopefully expect a considerable enlargement of the history of nature, without which our hypotheses are but dreams and romances, and our science meer conjecture and opinion.[72]

Sharing the general 17th-century contempt for antiquity, Glanvill suggested that prior to the modern era, no sound theory had ever been proposed because philosophers were not sufficiently empirical.[73] He was especially scornful of Aristotle's excursions into natural science: "the Aristotelian hypotheses give us a very dry and *jejune* account of nature's phenomena".[74] On the whole, Glanvill is satisfied with almost any hypothesis which saves the appearances; at least he criticizes only those hypotheses which are incompatible with experience. But if two hypotheses both accord with nature equally well, he suggests that we should believe the simpler of the two:

whichever doth with more ease and congruity solve the phenomena, that shall have my vote for the most philosophic hypothesis.[75]

Glanvill was not only pessimistic about the possibility of finding true hypotheses about unobservable mechanisms, he even doubted that we could confidently discuss causal relations between observable objects. In a passage that has led some scholars to rank him as the '17th-century Hume', he argues:

So that we cannot conclude, anything to be the cause of another; but from its continual accompanying it: for the causality itself is insensible. But now to argue from a concomitance to a causality is not infallibly conclusive.[76]

He goes on to insist that because we cannot "infallibly assure ourselves" of the truth of even the most obvious causal relation, "the foundation of scientifical procedure is too weak to permit us to build an undubitable science upon it".[77]

HENRY POWER

The final figure I want to consider is Henry Power (1623–1688), whose only claim to importance is based on the microscopical observations reported in his *Experimental Philosophy in three books: Containing new experiments – microscopical, mercurial, magnetical – with some deductions and probable hypotheses, raised from them in avouchment and illustration of the now famous Atomical Hypothesis* (1664). Like Boyle and Glanvill, Power pays homage equally to the "ever-to-be-admired Descartes"[78] and "the learned Verulam".[79] Conflating Descartes and Bacon much as Boyle and Glanvill did, Power is perfectly willing to allow hypotheses in science, so long as they "save all the appearances"[80] and are "confirmed and made good"[81] by several experiments, rather than a single one. But without experiments, he is convinced that "our best philosophers will prove but empty conjecturalists, and their profoundest speculations herein, but glossed outside fallacies."[82] Although not altogether opposed to hypotheses, Power is more optimistic than Descartes, Boyle, and Glanvill were about the possibility of achieving indubitable knowledge about the physical world. As we have seen, until Power's time, the method of hypothesis had invariably accompanied the corpuscular philosophy; for it seemed one could never do more than conjecture about the properties of the smallest bits of matter. But Power gives corpuscularism a new twist. Impressed by the microscope's ability to penetrate into the inner structure of organisms, Power suggests that by the aid of the microscope we will eventually be able to see corpuscles and "determine their precise mechanisms".[83] He envisions the erection of a "true and permanent philosophy" based on the microscopical canvassing of nature and the "infallible demonstrations of mechanicks".[84] Power believes that the microscope thus provides the key for breaking into Descartes' universal clock and out of his probabilism; for such an instrument offers us a glimpse of the internal mechanisms themselves. Power proceeds to turn the clock analogy against Descartes, suggesting that though the 'system-builders' can never observe more than the outer appearances, the faithful experimenter, with the help of optical instruments, will eventually learn the truth:

For the old dogmatists and notional speculators, that onely gazed at the visible effects and last resultances of things, understood no more of nature, than a rude countreyfellow does of the internal fabrick of a watch, that only sees the index and horary circle, and perchance hears the clock and alarum strike in it; But he that will give a satisfactory account of the phenomena, must be an artificer indeed, and one well-skilled in the wheel work and internal contrivance of such anatomical engines.[85]

After Power, writers like Hooke,[86] Newton,[87] and Cowley[88] enlarged on this theme, suggesting quite confidently that improved instrumentation promised hope of discovering the true mechanisms of nature and conclusively establishing (or refuting) the corpuscular philosophy.[89] As the faith in the unlimited magnifying powers of the microscope grew, the method of hypothesis waned, especially in England.[90] After all, the clock analogy is only persuasive when there is serious doubt about the possibility of perceiving nature's mechanisms. As such doubt faded, so did Descartes' influence and the method of hypothesis. Descartes thus fell victim to his own metaphor; victimized because the clock analogy, which he endorsed so enthusiastically, suggested to many later writers that the inner mechanisms of nature, like those of a clock, could eventually be directly scrutinized by careful observation and instrumentation. The demand for a hypothesis-free science, which was widely circulated after Newton, could never have gathered such enthusiastic adherents if the probabilism of Descartes, Boyle, and Glanvill had not died such a quick and needless death at the hands of those who thought nature's clock had no secrets which man's instruments could not seek out and know with certainty.[91] As it happened, the method of hypothesis went into virtual eclipse after 1700 until its revival almost a century later.[92]

NOTES

[1] Among the more important accounts of Cartesian influences in Britain, see: M. Nicolson, 'The Early Stage of Cartesianism in England', *Studies in Philology* **26** (1929), 356–74; J. Saveson, 'Descartes' Influence on John Smith, Cambridge Platonist', *J. Hist. Ideas* **20**, 255–63; S. Lamprecht, 'The Role of Descartes in 17th-Century England',

Studies in the History of Ideas, Boulder, Colorado, 1935; E. Burtt, *Metaphysical Foundations of Modern Science*, New York, 1932, *passim*; and Marie Boas [Hall], 'The Establishment of the Mechanical Philosophy', *Osiris* **10** (1952), 412–541.

2 For example, Jones asserts that "Experimental philosophy remains a thing distinct from the mechanical [and hypothetical] and Bacon, who was the chief sponsor of the former, far outweighs in importance Descartes, who lent his great influence to the latter ... Needless to say, the scientific movement in England in the third quarter of the seventeenth century ... was largely inspired by the great Chancellor [Bacon] ... " (*Ancients and Moderns*, St Louis, 1961, p. 169). Elsewhere he notes, "it is a mistake to think Cartesianism inspired the scientific movement in England" (ibid., p. 185). Jones goes so far as to suggest that this period in English science should be called the "Bacon-faced generation" (ibid., p. 237 ff.).

F. W. Westaway, another writer who denies Descartes' influence on English methodology, asserts that "Cartesianism took but slight hold in England" (*Scientific Method: Its Philosophy and Practise*, London, 1919, p. 127). With Boyle in particular, historians have been quick to apply the Baconian label. Thus, Butterfield, in a long discussion of Boyle's ideas, clings to the view that Boyle was a devout follower of Bacon, without hinting about a possible debt Boyle might owe to Descartes (cf. H. Butterfield, *Origins of Modern Science*, London, 1957, pp. 130–8). Marie Boas [Hall], argues that Boyle's corpuscularism (and the methodology which sustains it) was not derived from Descartes but was, rather, "an independent development along lines suggested by Bacon" (*Osiris* **10** (1952), 461). Recently, however, Hall has conceded that "though it was Bacon who mainly inspired Boyle, he was influenced by Descartes as well" (*Robert Boyle on Natural Philosophy*, Bloomington, Indiana, 1965, p. 63).

3 The work of scholars such as F. R. Johnson (*Astronomical Thought in Renaissance England*, Baltimore, 1937) makes it highly doubtful whether the experimental spirit of English science can be attributed to Bacon at all. Many of Bacon's predecessors and contemporaries (e.g., Harvey and Gilbert) were accomplished experimentalists long before the appearance of the *Novum Organum*.

4 Historians are gradually beginning to recognize the importance of Descartes' version of the method of hypothesis and the fundamental role it played in his philosophy of science. Especially useful in this regard are G. Buchdahl's discussions in 'Descartes' Anticipation of a Logic of Scientific Discovery', *Scientific Change* (ed. A. C. Crombie), London, 1962, pp. 399–417, and 'The Relevance of Descartes' Philosophy for Modern Philosophy of Science', *Brit. J. Hist. Sci.* **1** (1963), 277–49. See also R. Blake, 'The Role of Experience in Descartes' Theory of Method', in *Theories of Scientific Method* (ed. E. Madden), Seattle, 1960, pp. 75–103.

5 Although an English translation of the *Discourse* appeared in London in 1649, its circulation seems to have been quite limited. Apparently Descartes' *Passions of the Soul* was widely circulated in Britain, but since it has little of methodological interest, we shall neglect it in our discussion.

6 As he put it in the *Discourse*: "But I must confess also that the power of nature is so vast and ample, and these principles are so simple and general, that I observed hardly any particular effect concerning which I could not at once recognize that it might be deduced from the principles in many different ways and my greatest difficulty is usually to discover in which of these ways the effect does depend on them" (R. Descartes, *Philosophical Works*, trans., Haldane and Ross, New York, 1931, vol. i, p. 121).

[7] R. Descartes, *Oeuvres* (ed. Adam and Tannery), Paris, 1897–1957, vol. ii, p. 199.
[8] "I frankly confess that concerning corporeal things, I know only this: that they can be divided, shaped and moved in all sorts of ways . . . " (ibid., vol. ix, p. 102).
[9] Ibid., vol. ix, p. 322. The passage in square brackets occurs only in the French edition of the *Principles*, not the Latin.
[10] Ibid., vol. viii, p. 326.
[11] Recall Descartes' classic remark that: "As for physics I should believe myself to know nothing of it if I were only able to say how things may be, without demonstrating that they cannot be otherwise" (ibid., vol. iii, p. 39).
[12] In the *Dioptrics*, for example, Descartes tries to explain different colored rays of light. He suggests that light is composed of a pulse transmitted at infinite velocity by contiguous spherical corpuscles. These spheres have a rotational motion about their centers. Different colored rays of light are due to the differential speeds of axial rotation which the corpuscles can assume. A fast spin appears to be red light, a moderate spin as yellow, and a slow spin as blue. With this model, Descartes succeeds in explaining the phenomena of color without recourse to entities except matter-in-motion. But at the same time, he has been forced to go beyond the knowledge given by the first principles to hypothesize, with neither empirical evidence nor *a priori* reasons, that different atoms of matter rotate at different speeds and that such rotation is the cause of color. Thus, he deduces the phenomena of color from the conjunction of the first principles 'matter' and 'motion' *and* an assumption about differential speeds of rotation.
[13] G. Buchdahl in *Scientific Change* (ed. A. C. Crombie), London, 1962, p. 411.
[14] R. Descartes, *Oeuvres*, 1897–1957, vol. ix, p. 325.
[15] Ibid., vol. ii, p. 199.
[16] Ibid., vol. ix, p. 123.
[17] D. J. de Solla Price ("Automata and the Origins of Mechanism and Mechanistic Philosophy", *Technology and Culture* 5 (1964), 9–23) has delineated the fundamental role that clocks and other automata played as analogies for the mechanical and corpuscular scientists. However, Price has not drawn attention to the *methodological* ramifications of the clock-analogy which I intend to discuss in this chapter.
[18] I must make it as explicit as possible that this paper is *not* an attempt to date precisely the introduction of Cartesian ideas into English science and philosophy. For that, we should probably have to look closely at Hobbes, Digby, Charleton, Cudworth, and More rather than Boyle and the other writers I deal with. My goal is a rather different one, namely to suggest that there are certain Cartesian strains which loom large in Boyle's methodological writings. Whether they came directly from Descartes or through an intermediary source is a separate question which I touch on only incidentally. (Cf. L. Gysi, *Platonism and Cartesianism in the Philosophy of Ralph Cudworth*, Bern, 1962.)
[19] Boyle's unpublished manuscripts leave absolutely no doubt that he has read Descartes, and more than once. From his earliest papers on natural philosophy until his last ones, he made repeated references to Descartes and the Cartesians. There is one particularily interesting passage in which, discarding his normal humility, he candidly assesses the contributions of several important 17th-century scientific figures. His admiration for Descartes is certainly undisguised: "Hobbes est obscur sans agrément, singulier en ses idées, scavant, mais peu solide, inconstant dans sa doctrine: car il est tantost Epicurien, tantost Peripateticien. Boile [he evidently says of himself] est exact dans ses observations: il n'y a personne en l'Europe qui ait enrichy la philosophie de tant d'experiences

que luy: il raissone assez consequemment sur ses experiences, lesquelles aprés tout ne sont pas toujours indubitables: parce que ces principles ne sont pas toujours certains . . . Gassendi, qui n'a voulu passer que pour restauranteur de la philosophie de Democrite et d'Epicure, parle peu de son chef, il n'a presque rien de luy, que la beauté du stile, par ou il peut passer pour un auteur admirable: pour le refuter dans san physique, on n'a besoin que des argumens d'Aristote contre Democrite et ses disciples. Descartes est un genie des plus extraordinaires qui ait paru dans ces derniers temps, d'un esprit fertile, et d'une meditation profound: L'enchainement de sa doctrine va à son but, l'ordre en est bien imaginé, selon ses principes: et son systeme, tout melé qu'il est d'ancien et de moderne, est bien arrangé. A la verité il enseigne trop ă douter: et ce n'est pas un bon modele à des esprits naturallement incredules: mais enfin il est plus original que les autres . . . Enfin Galilei est le plus agreeable des modernes, Bacon le plus subtil, Gassendi le plus scavant, Hobbes le plus resveur, Boyle le plus curieux, Descartes le plus ingenieu, Vanhelmont le plus naturaliste: mais trop attaché a Paracelse" (*Royal Society, Boyle Papers*, vol. xliv).

20 Boyle was not the first to suggest the similarity between Bacon and Descartes. The anonymous translator of Descartes' *Passions of the Soule*, London, 1650, insists, in an 'advertisement' appended to that work, that though "most men conceive not how necessary experiments are", Descartes and Bacon "had the best notions, concerning the method to be held to bring the Physicks to their perfection". A good account of Boyle's scientific method is to be found in the second chapter of M. Mandelbaum, *Philosophy, Science, and Sense Perception*, Baltimore, 1964, esp. pp. 88–112.

One should also mention R. Westfall's useful 'Unpublished Boyle Papers Relating to Scientific Method', *Ann. Sci.* 12 (1956), 63–73 and 103–17; Marie Boas [Hall], 'La Méthodologie Scientifique de Robert Boyle', *Rev. Hist. Sci.* 9 (1956), 105–25; and A. R. and M. B. Hall, 'Philosophy and Natural Philosophy: Boyle and Spinoza', in *Mélanges Alexandre Koyré* (ed. I. B. Cohen and R. Taton), Paris, 1964, vol. ii, pp. 241–56. Older, but still useful, sources are S. Mendelssohn, *Robert Boyle als Philosoph*, Wurzburg, 1902 and G. Sprigg, 'The Honorable Robert Boyle: A Chapter in the Philosophy of Science', *Archeion* 11 (1929), 1–12.

21 The very fact that Boyle close to call his doctrine 'the *corpuscular* philosophy' is indicative of his filiations with Descartes. Prior to Boyle, it was common to distinguish three distinct theoretical systems: the Aristotelian, the Cartesian and the Atomic or Epicurean or Gassendian. Boyle rightly pointed out that such a classification obscured the considerable area of agreement between the Cartesian and Atomistic paradigms. Rather than call himself an atomist, and thereby side with Gassendi against Descartes, Boyle defines a more general position ('the corpuscular philosophy') which permits him to consider himself in a Cartesian tradition while still siding with the Atomists on many specific points of interpretation. He puts it this way: ' . . . I considered that the Atomical and Cartesian hypotheses, though they differed in some material points from one another, yet in opposition to the Peripatetic and other vulgar doctrines they might be looked upon as one philosophy . . . [for] both the Cartesians and the Atomists explicate the same phenomena by little bodies variously figured and moved . . . their hypotheses might by a person of reconciling disposition be looked upon as one philosophy" (*Works*, ed., Birch, London, 1772, vol. i, pp. 355–56).

Boyle unreservedly regards Descartes as the mechanical *cum* corpuscular philosopher *par excellence*: "That strict philosopher Descartes who has with great wit and no less applause attempted to carry the mechanicall powers higher than any of the modern philosophers and apply it to explicate things mechanically" (*Royal Society, Boyle Papers*, vol. ii, f. 137: cf. also *Works*, vol. iii, p. 558).

[22] *Royal Society, Boyle Papers*, vol. ix, f. 1.

[23] As T. S. Kuhn puts it: "Neither [Boyle's] eclecticism nor his scepticism extends to doubts that some corpuscular mechanisms underlies each inorganic phenomenon, he investigates" ('Robert Boyle and Structural Chemistry in the Seventeenth Century', *Isis* **43** (1952), 19).

[24] Robert Boyle, *Works*, 1772, vol. ii, p. 45.

[25] *Royal Society, Boyle Papers*, vol. ix, f. 40. Elsewhere he makes the point this way: it would "be backward to reject or despise all explications that are not immediately deduced from the shape, bigness and motion of atoms or other insensible particles of matter . . . [for those who] pretend to explicate every phenomenon by deducing it from the mechanical affections of atoms undertake a harder task then they imagine" (ibid., vol. viii, f. 166).

[26] Ibid., vol. viii, f. 184. Underlined in original. Cf. note 63 below.

[27] R. Boyle, *Works*, 1772, vol. iv, p. 72.

[28] *Royal Society, Boyle Papers*, vol. ix, f. 40.

[29] Boyle even suggests that the subordinate hypotheses may be the only ones which can be firmly established: "Though men be not arrived at such a pitch of knowledge as to be able to discover and solemnely-established [sic] compleat and *general hypotheses*; yet subordinate *axioms* and *hypotheses* . . . may be of vast use both in philosophy and to human life" (ibid., vol. ix, f. 61).

[30] *Royal Society, Boyle Papers*, vol. ix, f. 28.

[31] R. Boyle, *Works*, 1772, vol. v, pp. 513–15.

[32] Boyle was as dogged as Bacon in putting histories of nature high on his list of priorities: " . . . we evidently want that upon which a theory, to be solid and useful, must be built; I mean an experimental history . . . And this we so want, that except perhaps what mathematicians have done concerning sounds, and the observations (rather than experiments) that our illustrious Verulam hath (in some few pages) said of heat in his short *Essay de Forma Calidi*; I know not any one quality of which any author has given us an anything competent history" (ibid., vol. iii, p. 12).

[33] Birch, the editor of Boyle's *Works*, records no usage of "induction among the writings contained in those volumes". Boyle's unpublished MSS. in the Royal Society Library are not indexed.

[34] I do not mean to suggest by this that Bacon's methodology left no room for the hypothetical method or that it was the invention of Descartes. To the contrary, I believe Bacon's 'Indulgence of the Understanding or First Vintage' is an unequivocal statement of the hypothetical method. However, I think that the myth of Bacon the anti-hypotheticalist has this much truth in it: that Bacon's Hypothetical Indulgence of the Understanding was conceived by him as a strictly temporary measure until the natural histories became sufficiently complete to permit the fool-proof mechanical induction which Bacon outlines in the *Novum Organum*. Bacon was scornful of the view (supported by both Descartes and Boyle) that science must be forever conjectural and uncertain. Furthermore, Bacon leaves completely unmentioned the particulars of his First Vintage and so I think it is to Descartes, rather than Bacon, that Boyle turned for the details of his methodology. Bacon could never have accepted, as Boyle enthusiastically did,

Descartes' clock-maker analogy and its implications for the severe limitations on scientific knowledge.

35 *Royal Society, Boyle Papers*, vol. ix, f. 105. Boyle tells us that one of the primary functions of experimentation is "to suggest hypotheses" (ibid., f. 30).

36 R. Boyle, *Works*, 1772, vol. iv, p. 234.

37 Boyle is quite blunt in his criticism of the view that experiment alone will lead us to true theories: "he, that establishes a theory, which he expects shall be acquiesced in by all succeeding times ... must not only have a care, that none of the phenomena of nature, that are already taken notice of, do contradict his hypothesis at the present [time], but that no phenomena, that may be hereafter discovered, shall do it for the future." But considering "how incompleat the history of nature we yet have is, and how difficult it is to build an accurate hypothesis upon an incomplete history of the phenomena", we can never say with certainty that our theories are true (ibid., vol. iv, p. 59).

38 Boyle writes: "it is sometimes conducive to the discovery of truth, to permit the understanding to make an hypothesis, in order to the explication of this or that difficulty, that by examining how far the phenomena are, or are not, capable of being solved by that hypothesis, the understanding may, even by its own errors, be instructed" (ibid., vol. i, p. 303). This statement, along with several others, could be cited as counter-evidence to Jones' assertion that "Boyle was influenced by the comprehensive characteristic of Bacon's philosophy in thinking that all the evidence must be in before a generalization should be drawn" (op. cit., footnote 2, p. 164). Boyle clearly and Bacon dimly, perceived the necessity of making hypotheses and generalizations before *all* the evidence was collected.

39 R. Boyle, *Works*, 1772, vol. ii, p. 45. Boyle formulates the clock analogy on several occasions, which suggests that it played a basic role in his thinking about science and method. This conjecture is confirmed when Boyle says, in introducing the clock metaphor: "To explain this a little, let us assume the often mentioned and often to be mentioned, instance of a clock" (*Royal Society, Boyle Papers*, vol. ii, f. 141).

40 R. Boyle, *Works*, 1772, vol. ii, p. 45. Compare this with Descartes' remark that "one may deduce some very true and certain conclusions from suppositions that are false or uncertain" (*Oeuvres*, 1897–1957, vol. ii, p. 199).

41 Cf. *Works*, 1772, vol. ii, pp. 46 ff.

42 Ibid., vol. 1, p. 82. Cf. also *Royal Society, Boyle Papers*, vol. ix, f. 63. In principle 205 of the fourth book of the *Principles*, Descartes had similarly compared scientific theorizing with decoding techniques: "If, for instance, anyone wanting to read a letter written in Latin characters not in their proper order, decides to read B wherever he finds A and C where he finds B ... and if he in this way finds there are certain Latin words composed of these, he will not doubt that the true meaning is contained in the words, although he discovered this by conjecture, and although it is possible that the writer did not arrange the letters in this order of succession, but in some other ... " (*Oeuvres*, 1897–1957, vol. ix, p. 323). Cf. also Rule 10 of the *Regulae*.

43 "And lest it be supposed that Aristotle did, or wanted to do more than this, it must be recalled that he expressly says in the first book of the *Meteors*, at the beginning of the seventh chapter, that with regard to things not evident to the senses, he thinks that he offers sufficient explications and demonstrations of them, if he merely shows that they may be as he explains them" (ibid.).

[44] R. Boyle, *Works*, 1772, vol. ii, p. 45. It is significant that another English proponent of atomism, Walter Charleton, who similarly adopted a hypothetical theory of science (our conjectures tell us how the world "*may* be, rather than how it is or must be", *Physiologica*, London, 1654, p. 128), also appeals to Aristotle's remark in the *Meteors*, to support his hypotheticalism (ibid.). Immediately thereafter, he quotes Descartes' *Principles* on the same point. This raises the possibility that Boyle borrowed the reference to Aristotle's *Meteors* from Charleton rather than directly from Descartes. However, Boyle states that the portion of his MSS. in which the reference to Aristotle occurs was written in 1651 of 1652, *prior* to the appearance of Charleton's *Physiologica*. (cf. note 91). Although Boyle may have misremembered the date of composition, there is no evidence that he has done so. (For a brief, but suggestive account of the relations between Boyle and Charleton, see R. Kargon, 'Walter Charleton, Robert Boyle and the Acceptance of Epicurean Atomism in England', *Isis* 55 (1964), 184–92.)

[45] R. Boyle, *Works*, 1772, vol. ii, p. 37.

[46] Ibid., vol. iv, p. 234.

[47] "Although there exist several individual effects to which it is easy to adjust diverse causes [i.e., hypotheses] one to each, it is however not so easy to adjust one and the same [hypothesis] to several different effects, unless it be the true one from which they proceed" (R. Descartes, *Oeuvres*, 1897–1957, vol. ii, p. 198).

[48] This principle received its definitive formulation at Newton's hands, in his third Rule of Philosophizing.

[49] R. Descartes, *Oeuvres*, 1897–1957, vol. ix, p. 319.

[50] R. Boyle, *Works*, 1772, vol. iv, p. 72. For a more extensive discussion of the principle of multi-level identity in the 17th century, see Chapter 5.

[51] I have drawn, in part, from the unpublished Boyle material brought to light by R. Westfall (op. cit., see note 20, above). However, Westfall gives no consideration to the Cartesian origins of Boyle's hypotheticalism, suggesting rather the similarities between Bacon and Boyle.

[52] R. Boyle, *Works*, 1772, vol. i, pp. 302–303. The latter part of this passage confirms a point we suggested earlier, viz., that Boyle concieved his hypothetical method as perfectly compatible with Bacon's Indulgence of the Understanding. In making the point, Boyle even adopts Bacon's language ("it is sometimes conducive . . . to permit the understanding to make an hypothesis . . . ").

[53] Ibid., p. 303.

[54] *Royal Society Boyle Papers*, vol. ix, f. 113.

[55] Hobbes' probabilistic account of scientific knowledge is very much like Descartes', even in language. For instance, Hobbes asserts: " . . . he that supposing some one or more motions, can derive from them the necessity of that effect whose cause is required, has done all that is to be expected from natural reason. And though he prove not that the thing was thus produced, yet he proves that thus it may be produced . . . which is as useful as if the causes themselves were known" (*English Works*, ed. W. Molesworth, London, 1845, vol. vii, pp. 3–4). Compare this with the latter half of the passage in which Descartes formulates the clock analogy.

[56] R. Boyle, *Works*, 1772, vol. i, p. 241. For an account of Hobbes' views, see E. Madden, 'Thomas Hobbes and the Rationalistic Ideal', in *Theories of Scientific Method* (ed. Madden), Seattle, 1960, pp. 104–18.

[57] R. Boyle, *Works*, 1772, vol. iv, p. 77.

58 R. Westfall, op. cit. (note 20), pp. 113–14.

59 M. B. Hall, *Robert Boyle on Natural Philosophy*, Bloomington, Indiana, 1965, pp. 134–5.

60 *Royal Society Boyle Papers*, vol. ix, f. 25.

61 In light of the analysis offered here, it is surprising to find M. Cranston writing that "Boyle's inductive method was essentially Baconian" (*John Locke*, London, 1957, p. 75). Cranston is wrong on two accounts; Boyle's method was neither inductive nor Baconian.

62 R. Harré has made a similar point about the connexions between 17th-century metaphysics and methodology, focussing especially on the links between the corpuscular philosophy and the doctrine of primary-secondary qualities (*Matter and Method*, London, 1964). However, Harré's analysis is handicapped by his assumption that Descartes had nothing to do with either the metaphysics or the methodology or corpuscularism.

63 It is thus unjustified to juxtapose Boyle and Descartes as if they personified, respectively, empiricism and rationalism. A typical instance of such juxtaposition is to be found in a recent essay by A. R. and M. B. Hall. They write: "Is a scientific proposition demonstrated when it is shown to be a logical consequence of a set of intuitively certain axioms? Descartes would have answered affirmatively, Boyle negatively. Is a scientific proposition sufficiently demonstrated only by showing empirically that it holds? Descartes would have answered negatively, Boyle affirmatively" ('Philosophy and Natural Philosophy: Boyle and Spinoza', in *Mélanges Alexandre Koyré* ed. I. B. Cohen and R. Taton, Paris, 1964, vol. ii, pp. 242–243). From the evidence presented here, one could more reasonably conclude that Boyle and Descartes would both answer each of the two questions affirmatively.

Boyle is so far from being the strict empiricist the Halls suggest he is that he occasionally adopts a rationalism more rigorous than Descartes'. For example, he observes that "it is not always necessary, *though it be always desirable*, that he, that propounds an hypothesis in astronomy, chemistry, anatomy, or other part of physics, be able *a priori*, to prove his hypothesis to be true . . . " (*Works*, 1772, vol. iv, p. 77). Cf. footnote 29 above. Elsewhere, and again in a most 'unempirical' vein, Boyle states that "where reason proceeds in a due manner . . . its conclusions are to be preferred to some testimonies of sense" (*Royal Society, Boyle Papers*, vol. ix, f. 33).

But it would equally be mistaken to think that Boyle was altogether clear in his own mind about the precise relations between reason and experiment; for there certainly are times when Boyle opts for the position traditionally called empiricism. At one point, in a startling anticipation of Newton's Fourth Rule of Philosophizing, Boyle says "that the well circumstanc'd testimony of sense is to be preferred to any hypothesis . . . " (ibid., vol. ix, f. 31).

64 For a general account of Glanvill's scientific work, see M. Prior's 'Joseph Glanvill, Witchcraft and Seventeenth-Century Science', *Modern Philology* **30** (1932), 167–93.

65 J. Glanvill, *Scepsis Scientifica* (ed. Owen), London, 1885, p. 44.

66 Ibid., p. 155. This same passage occurs in his earlier *Vanity of Dogmatizing*, London, 1661, p. 180.

67 J. Glanvill, *Scepsis Scientifica*, London, 1885, p. 44.

68 Ibid., pp. 182–3.

69 Even in the early pages of the *Scepsis Scientifica*, Glanvill is pre-occupied with the clock analogy which he makes explicit in the passage cited above. Throughout his work,

the model of nature as an observable but unknowable machine floats just beneath the surface: "We cannot profound into the hidden things of nature, nor see the first springs and wheels that set the rest a going. We view but small pieces of the universal frame, and want phenomena to make intire and secure hypotheses" (ibid., p. 75).

70 Ibid., pp. 170–1.

71 Glanvill indicates his sympathy for the English polemic against systems when he writes: "if such great and instructed spirits [i.e, the members of the Royal Society] think we have not as yet phenomena enough to make as much as an hypothesis; much less to fix certain [i.e., indubitable] laws and prescribe method to nature in her actings: what insolence is it then in the lesser size of mortals, who possibly know nothing but what they have gleaned from some little *systeme* . . . to boast infallibility of knowledge . . . " (ibid., p. li).

72 Ibid., p. lxii.

73 " . . . 'tis possible that all the hypotheses that have yet been contrived, were built upon too narrow an inspection of things . . . " (ibid., p. lxiii).

74 Ibid., p. 145.

75 Ibid., p. 51.

76 Ibid., p. 166.

77 Ibid., pp. 167–8.

78 H. Power, *Experimental Philosophy* . . . London, 1664, preface.

79 Ibid.

80 Ibid., p. 94.

81 Ibid., p. 114.

82 Ibid., preface.

83 Cf. Ibid., p. 82.

84 Ibid., p. 192.

85 Ibid., p. 193. In his *Essay*, John Locke likens our knowledge of sub-microscopic mechanisms to a "countryman's idea" of the "inward contrivance of that famous clock at Strasbourg, whereof he only sees the outward figures and motions" (*Essay*, London, 1929, iii, 6, para. 9). Could this be a gloss on Power's formulation of the clock-maker analogy? For a discussion of Locke's general methodological position, see R. Yost, 'Locke's Rejection of Hypotheses About Sub-Microscopic Events', *J. Hist. Ideas* **12** (1951), 111–30, and chapter 5.

86 Hooke writes: " . . . and by the help of *microscopes*, there is nothing so *small* as to escape our inquiry . . . It seems not improbable but that by these helps [viz., optical instruments] the subtilty of the composition of bodies, the structure of their parts, the various texture of their matter, the instruments and manner of their inward motions, and all the possible appearances of things, we may come to be more full discovered . . . " (*Micrographia*, London, 1665, preface). What we cannot discover about corpuscles by seeing them, Hooke suggests we can learn by listening to them: 'There may also be a possibility of discovering the internal motions and actions of bodies by the sound they make, who knows but that as in a watch we may hear the beating of the balance, and the running of the wheels, and the striking of the hammers, and the grating of the teeth, and the motions of the internal parts of bodies, whether animal, vegetable, or mineral by the sound they make . . . " (*Posthumous Works* (ed. Waller), London, 1705, p. 39). Considering that both Hooke and Power looked to instrumentation to free science from its dependence on hypotheses, it is perhaps significant that in the preface to his

Micrographia, Hooke mentions Power's *Experimental Philosophy*. He goes on to remark that he and Power examined one another's MSS. before they went to press.

87 In the *Opticks*, Newton exudes a similar optimism when he writes that: "It is not impossible but that microscopes may at length be improved to the discovery of the particles of bodies on which their colours depend, if they are not already in some measure arrived to that degree of perfection" (*Opticks*, New York, 1952, fourth edition, p. 261). Optics, on Newton's view, is likely to become a completely inductive, observational, and non-hypothetical science. (Cf. R. Kargon, 'Newton, Barow and the Hypothetical Physics', *Centaurus* 11 (1965), 46–56, for an account of Newton's reaction to the Cartesian method of hypothesis. It is interesting to compare Newton's optimism about seeing corpuscles with Charleton's remark that even the atoms "of the largest size, or rate, are much below the perception and discernment of the acutest optics, and remain commensurable only by the finer digits of rational conjecture" (op. cit., p. 113).

88 In his *Ode to the Royal Society*, Cowley gave poetic expression to Power's optimistic hopes for the microscope giving us infallible knowledge of the inner parts of nature's clock:

> "Nature's great works no distance can obscure,
> No smallness her near objects can secure.
> You've taught the curious sight to press
> Into the privatest recess
> Of her imperceptible littleness . . .
> You've learned to read her smallest hand,
> And well begun her deepest sense to understand."

(In Sprat's *History of the Royal Society*, London, 1667, immediately following the 'epistle Dedicatory'.)

89 If we strain the point a little, we can see influences of the clock analogy stretching even into the early 18th century. In his preface to the second edition of Newton's *Principia*, Roger Cotes – in a vein very much like Henry Power – turns the analogy against the Cartesians. After a vigorous attack on the hypotheticalism of the Cartesians, Cotes insists that: 'The business of true philosophy is to derive the natures of things from causes truly existent, and to inquire after those laws on which the Great Creator actually chose to found this most beautiful frame of the world, not those by which he might have done the same, had he so pleased . . . The same motion of the hour-hand in a clock may be occasioned either by a weight hung or a spring shut up within. But if a certain clock should be really moved with a weight, we should laugh at a man that would suppose it moved by a spring . . . for certainly the way he ought to have taken would have seen actually to look into the inward parts of the machine, that he might find the true principle of the proposed motion' (*Principia*, 4th ed., Berkeley, 1934, pp. xxvii–xxviii). Descartes' point that the inner machinations of nature – the connexions between phenomena – are necessarily precluded from view is apparently lost on the optimistic Cotes.

90 Although many scientists evidently believed the microscope pointed the way to complete observability of micro-mechanisms, there were others who, seeing the real substance of Descartes' argument, took precisely the opposite position. Jacques Rohault, for instance, argued in the *Traité de Physique*, (Paris, 1671), that the moral to be drawn from the new world which the microscope unfolds is not that we are finally getting down

to ultimate reality, but rather, that nature is infinitely complex – that each new stage of magnification will reveal still finer and more subtle processes. Though the microscope permits us to examine the fleas on a dog, it does not permit us to see the micro-organisms which infest the flea. For Rohault, there is a potentially infinite regress to ever-smaller bits of matter which the microscope, however great its magnification, is powerless to exhaust.

[91] Considering that Boyle, Glanvill and Power all used the clock analogy in publications between 1661 and 1664, one is inclined to ask whether they borrowed the analogy from Descartes or from one another. I have been able to find little evidence which would conclusively resolve this antiquarian puzzle. In point of publication, Glanvill's *Vanity of Dogmatizing* (1661) was "so scarce as to be hardly known at all except by name" (*Scepsis Scientifica*, p. xvi). Furthermore, Boyle's *Usefulness of Natural Philosophy*, in which he used the clock analogy, though published in 1663, was sent to press in "1660, or 1661, and 1663" (Boyle, *Works*, vol. ii, p. 3) and Boyle claims that most of the first part of that work – in which the crucial passage occurs – was written ten or twelve years earlier (1651–53). In addition to the similarity of language between Boyle's and Descartes' formulations of the analogy (which would undermine the Descartes-to-Glanvill-to-Boyle view), we might note that Boyle never mentions Glanvill in his published works. As for Power, he never mentions Glanvill though he continually writes of Boyle and Descartes. Since Boyle's *Usefulness* predates Power's *Experimental History* (1664), it is possible that Power got the analogy from Boyle rather than Descartes. On the other hand, Power's book received the *imprimatur* in 1663 (when Boyle's *Usefulness* was barely off the press) and its preface is dated 1661. In light of such evidence (scanty though it is), I think it probable that all three writers got the analogy directly from Descartes rather than from one another.

[92] For an account of the revival of the method of hypothesis, see Chapter 8 below.

CHAPTER 5

JOHN LOCKE ON HYPOTHESES: PLACING THE *ESSAY* IN THE 'SCIENTIFIC TRADITION'

LOCKE ON HYPOTHESES

It has often been assumed by advocates of the purist view of the theory of knowledge (a view outlined in Chapter 2) that John Locke was primarily an epistemologist with only a casual and superficial interest in the physical sciences. Despite the fact that he studied medicine and spoke glowingly of figures like Newton and Boyle,[1] Locke's *Essay* seems – at least on the surface – to be concerned with the epistemology of commonsense rather than with the logic and methods of science. Philosophers, evidently by reading history backwards, have written as if Locke accepted the view of Berkeley and Hume that the empiricist philosophy should not base itself on a 'scientific' metaphysics. Furthermore, in so far as the *Essay* does deal with scientific matters, it usually seems to treat them with derision and condescension. Consequently, some commentators have inferred that Locke was an opponent of the corpuscular micro-physics which dominated the science of his day and have viewed his *Essay* as an attempt to develop a theory of knowledge with no corpuscularian, or other quasi-scientific, bias. They suggest that Locke was opposed not only to the atomic hypothesis, but to the use of virtually all hypotheses about imperceptibles in science. Those commentators who do not explicitly attribute to Locke a hostility to hypotheses generally leave unmentioned his remarks about scientific method, as if philosophy of science was foreign to the spirit of the *Essay*.[2] Recently, however, Maurice Mandelbaum has pointed out not only that Locke was sympathetic to the corpuscular program, but that an atomic view of nature is essential to Locke's epistemology and metaphysics.[3] Rather than read Locke in the light of Berkeley's criticisms, Mandelbaum urges us to approach the *Essay* with the atomic theories of Boyle and Newton in mind.

In this chapter, I want to build upon Mandelbaum's analysis by looking carefully at the theory of scientific method implicit in the *Essay*. For if Mandelbaum is right that Locke was vitally concerned with corpuscular physics, then we have every reason to expect that the *Essay* will provide guidelines for the way in which Locke – a life-long scientist himself – wanted to see science develop. Although it is hoped that this chapter will substantiate

Mandelbaum's reading of Locke as a corpuscularian, from a slightly different point of view, its primary aims are to ascertain Locke's attitude on the role of hypotheses in science and to use that attitude for 'locating' Locke within the contemporary tradition of scientific epistemology. The bearing of this latter problem on Mandelbaum's thesis should be clear; if Locke was as opposed to the method of hypothesis as most writers have made him out to be, then he could not conceivably have embraced so hypothetical a theory as the atomic one. Conversely, if Locke was sympathetic to the use of hypotheses in physics, then it would not be surprising if he adopted the corpuscular philosophy as enthusiastically as Mandelbaum maintains he did.

Perhaps the best place to begin is in response to one of the most detailed studies of Locke's methodology; namely, that of R. M. Yost.[4] In his lengthy analysis of the methodology of the *Essay*, Yost comes to the conclusion that Locke was not only sceptical about the scientific value of the atomic philosophy, but that he objected – on *methodological* grounds – to all scientific theories which employed hypotheses about unobservable events or objects. More specifically, Yost claims that "unlike many scientists and philosophers of the 17th century, Locke did not believe that the employment of hypotheses about sub-microscopic events would accelerate the acquisition of empirical knowledge".[5] Yost insists that while Locke allowed, albeit grudgingly, the use of hypotheses about observable events, he was categorically opposed to all hypotheses dealing with the behavior and properties of unobservable forces and atoms. He contrasts Locke's views on this subject with those of such 17th-century atomists as Boyle and Descartes, who (as urged above) endorsed the use of hypotheses about unobservable entities. Yost suggests that Locke's hostility to hypotheses about imperceptibles was a radical departure from the method of hypothesis which accompanied the atomism of his contemporaries. But Yost's analysis seems to overlook many of Locke's crucial pronouncements on methodology and to obscure the meaning of others. For not only did Locke look favorably on many uses of the method of hypothesis, but in doing so he was solidly in, rather than aligned against, the corpuscularian tradition. To develop this argument, I shall work in two directions. To begin with, I want to determine what Locke's attitude towards sub-microscopic hypotheses was. I shall then turn to consider his debt to the corpuscular philosophers who preceded and influenced him.

The casual reader of Locke's *Essay* comes away from that volume with the firm suspicion that Locke was uniformly pessimistic about the natural sciences. Apart from the general scepticism which forms the dominant motif of the fourth book of the *Essay*, there are numerous specific passages which reinforce this impression. Thus Locke asserts:

> As to a perfect science of natural bodies, . . . we are, I think, so far from being capable of any such thing, that I conclude it lost labour to seek after it.[6]

If it is ill-conceived even to attempt to develop a science of mechanics, how hopeless must the situation be for other branches of scientific inquiry. Elsewhere, he writes that 'scientifical' knowledge of nature is forever out of our reach.[7] Or again, he sadly proclaims that mechanics must not "pretend to certainty and demonstration", and that there can never be a "science of bodies" (Ibid.). Locke's pessimism seems especially pronounced whenever he discusses the corpuscularian program for explaining the observable world in terms of the motion and concretions of unobservable atoms. Thus,

> Because the active and passive powers of bodies and their ways of operating, consisting in a texture and motion of parts which we cannot by any means discover . . . .[8]

Or,

> I doubt not but if we could discover the figure, size, texture, and motion of the minute constituent parts of any two bodies, we should know *without trial* several of their operations upon one another . . . . But whilst we are destitute of senses acute enough [to perceive such corpuscles] . . . we must be content to be ignorant of their properties and ways of operations.[9]

Passages like these have disposed several historians of philosophy to interpret Locke both as an opponent of hypotheses in general and of mechanical hypotheses in particular. But Locke was neither of these. To see the flaws in caricaturing Locke as an opponent of the method of hypothesis, we need only recall that he devoted an entire section of the fourth book of the *Essay* (iv. 12. 13) to the "true use of hypotheses" and that he frequently spoke as if the phenomena of the visible world ultimately derived from interactions on the corpuscular level.[10] But if such remarks indicate Locke's acceptance of the method of hypothesis, we are confronted, when we compare them with *obiter dicta* like those cited above, with an obvious tension between Locke's apparent denunciation and acceptance of hypotheses.

Much of the apparent contradiction vanishes if we recall Locke's pivotal distinction between knowledge and judgment. Knowledge, for him, is based on a true and infallible intuition of the relation of ideas. To know that a statement '*x*' is true is to recognize that we could not conceive things to be other than the state of affairs which '*x*' specifies. In this way, we 'know' the truths of mathematics. But we do not 'know' anything about the physical world. Many statements the scientist makes may be highly probable, but they are not indubitably true and, because of this deficiency, are not in the domain of knowledge. When Locke says that we cannot 'know' anything about the "minute parts of bodies", he is using 'knowledge' in this technical sense. Since science is the name given to the body of our 'knowledge', natural philosophy can never be "scientifical."[11] But Locke was not so imprudent as to restrict our discourse rigidly to strictly "scientific" statements.[12] He recognized clearly that one can say some informative and highly probable things about the physical world. Such statements belong, however, not to knowledge, but to *judgment*:

> The faculty which God has given man to supply the want of clear and certain knowledge, in cases where that cannot be had is *judgment*: whereby the mind takes . . . any proposition, to be true or false, without perceiving a demonstrative evidence in the proofs.[13]

The natural philosopher may be able to make many very likely statements but "the highest probability amounts not to certainty; without which there can be no true knowledge" (Ibid., iv, 3, para. 14). Knowledge, then, consists in those statements which are clearly and distinctly perceived to be true; judgment consists of all those statements which are merely probable or conjectural: "Judgment is the presuming things to be so, without perceiving it."[14]

Having said that judgment deals with probable statements, Locke proceeds to argue that there are two sorts of probable statements: (1) those dealing with strictly observable phenomena, or "matters of fact", and (2) speculations dealing with unobservable phenomena.[15] Locke then turns his attention to the second sort of probable statement, viz., speculations about unobservables. Further sub-dividing the domain, he argues that there are two types of such speculations: (1) conjectures about purely spiritual beings (e.g., angels, demons, etc.,) and (2) hypotheses about the unobservable causes of such natural phenomena as generation, magnetism, and heat.[16] Locke's remarks on this second class are particularly of interest to our argument.

Suppose, Locke reasons, that we want to understand the ultimate nature of heat. Because we do not clearly perceive, or even dimly observe, the causes of heat, we cannot claim to 'know' anything about it.[17] All we can hope to offer are probable statements about its nature and causes. Since Locke believes that any tentative explanation of heat will be couched in terms of the behavior of unobservable corpuscles, he insists that observation can tell us nothing *directly* about the behavior of the heat-producing atoms. How then can we formulate any useful hypotheses at all? Locke's answer is straightforward: by conceiving submicroscopic corpuscles on *analogy* with bodies which we do perceive, viz., we must picture the smallest particles as miniature instantiations of the gross objects of perception. Indeed, "in things which sense cannot discover, analogy is the great rule of probability" (Ibid.). In the case of heat, the analogy we should make is obvious:

> Thus, observing that the bare rubbing of two bodies violently upon one another, produces heat, and very often fire itself, we have reason to think that what we call *heat* and *fire*, consists in a violent agitation of the imperceptible minute parts of the burning matter (Ibid.).

We must resort to analogies and models in conceiving the nature of submicroscopic events because, since our conjectures about them cannot be directly verified, the only reason we have for believing them to be even probable is that "they more or less agree to truths that are established in our minds" and because such conjectures are at least compatible with "other parts of our knowledge and observation" (Ibid.). "Analogy," he notes, "in these matters [viz., relating to unobservable events] is the only help we have, and it is from that alone we draw all ground of probability" (Ibid.).

It was a basic tenet of Lockean epistemology that all our ideas of external objects derive from sensation. Thus, it was perfectly natural for him to insist that our ideas about unobservable corpuscles must be based on, and derived from, ideas which visible bodies impress on our senses.

These remarks about analogy-based hypotheses dealing with submicroscopic events represent more than a grudging concession by Locke to his scientific colleagues. Indeed, he insists that the enunciation of analogical hypotheses is the most productive and theoretically fertile method which the sciences possess:

> This sort of probability, which is the best conduct of rational experiments, and the rise of hypothesis, has also its use and influence; and a wary reasoning from analogy leads us often into the discovery of truths and useful productions, which would otherwise lie concealed.[18]

Having said as much, we can see how misleading Yost's remark is that Locke's "method employed hypotheses, but they were hypotheses about correlations of observable qualities and did not refer to submicroscopic mechanisms".[19] Equally untenable is Yost's view that Locke "thought that one could infer *something* about submicroscopic mechanisms, but not enough to help us in making discoveries".[20] In the passage cited above, Locke explicitly states that the use of corpuscular hypotheses could lead us "into the discovery of truths, and useful productions". Elsewhere, he writes that "hypotheses, if they are well made, are at least great helps to the memory, and often direct us to new discoveries".[21] Yet another misleading claim is Yost's assertion that Locke never mentioned the hypothetical method "whenever he spoke of the methods of increasing empirical knowledge".[22] The same passage cited above stands as an obvious counterexample to this statement.

In enunciating his hypothetical account of science, Locke likened nature to a clock whose external appearances (e.g., hands moving, wheels grinding, etc.) are visible but whose internal mechanisms are forever excluded from view. The scientists' conception of nature is even "more remote from the true internal constitution" of the physical world than a "countryman's idea from the inward contrivance of that famous clock at Strasburg, whereof he only sees the outward figures and motions".[23] If we knew the "mechanical affections of the particles" of bodies, "as a watchmaker does those of a watch",[24] then we would not need to make hypotheses, but could have infallible, first-hand knowledge of nature's mechanisms. But because we can never get inside of nature's clock, we must be content to hypothesize about the possible arrangements of its parts on the basis of its external cortex.

Though Locke believed that scientific explanations should be based on hypotheses about corpuscular events, he insisted that the scientist should be circumspect in his use of such hypotheses. Hypotheses must never be called "principles", because such an honorific title makes them sound more trustworthy than they are.[25] Furthermore, we should never accept an hypothesis unless we have carefully examined the phenomena which it is designed to explain, and even then only if the hypothesis saves all the phenomena efficaciously. Locke is quite sensitive to the dangers of an unbridled hypothetical method and he often warns us against its excesses. Thus, one of the major sources of error which he cites is the clinging tenaciously to preconceived hypotheses and prejudging the facts on the basis of those hypotheses.[26]

But it is a mistake to say that Locke's critique of the extravagant exaggerations of the method of hypothesis indicates his aversion to all forms of that method when applied to unobservable entities. At one point, Locke explictly acknowledges that his animadversions upon hypotheses were not designed to preclude the scientist from hypothesizing, but only to make him wary about it:

> But my meaning is, that we should not take up any one hypothesis too hastily (which the mind, that would always penetrate into the causes of things, and have principles to rest on, is very apt to do), till we have very well examined particulars, and made several experiments, in that thing which we would explain by our hypothesis, and see whether it will agree to them all; whether our principles will carry us quite through, and not be

as inconsistent with one phenomenon of nature, as they seem to accommodate and explain another.[27]

The traditional account of Lockean methodology is certainly correct in its insistence that Locke believed we could not have *knowledge* about unobservable events; it is equally true that Locke was exceedingly scornful of those who believed one could make indubitable statements about the properties of unobservable corpuscles. But he was not opposed to the use of atomic hypotheses – or other hypotheses which invoked unobservable entities – so long as they made sense of the phenomena and were treated as merely probable judgments.

LOCKE AND THE TRADITION OF HYPOTHESIS

We may conclude from the foregoing discussion that Locke was neither opposed to hypotheses (if properly conceived) nor an adversary of the corpuscularians who used hypotheses about submicroscopic events. Indeed, so far as one can judge from the texts, Locke enthusiastically accepted the view that changes in the observable world are caused by, and explicable in terms of, changes on the atomic level.[28] I now want to suggest that the major features of Locke's method of hypothesis as well as many of the epistemological arguments whereby he justifies that method, are derived from, or at best are variations on, the methodological ideas of his corpuscularian contemporaries and predecessors. In particular, I shall claim that Locke probably derived the following methodological ideas from his corpuscularian contemporaries: (1) the insistence upon the provisional and tentative character of *all* scientific theories, (2) the view of nature as a clock whose internal mechanisms are not susceptible of direct analysis or observation, (3) the doctrine that hypotheses must be constructed on analogy with the behavior of observable bodies, and (4) the insistence, related to (3), that the hypotheses about imperceptible events must be compatible with laws of nature and phenomena other than those which they were devised to explain.

(1) Though more than willing to urge the use of hypotheses, Locke was apprehensive lest merely probable conjectures be taken for immutable truths. He emphasized that hypotheses ought not be called "principles" because such a linguistic convention might make us "receive that for an unquestionable truth, which is really, at best, but a very doubtful conjecture".[29] As we have seen, he went to some lengths to stress the tentative and hypothetical character of scientific inquiry. Unlike those who insisted that natural philosophy

must be infallible, Locke was more than willing to allow for a large degree of uncertainty. In this, he was of one mind with the corpuscular and mechanical philosophers. Robert Boyle, for example, wrote that scientific theories should be "looked upon only as temporary ones; which are not entirely to be acquiesced in, as absolutely perfect, or incapable of improving alterations."[30] In a similar vein, Hooke warns against treating any scientific theories as if they were indubitable:

If therefore the reader expects from me any infallible deductions, or certainty of *axioms*, I am to say for myself that those stronger works of wit and imagination are above my weak abilities ... Wherever he finds that I have ventur'd at any small conjectures, at the causes of the things I have observed, I beseech him to look upon them as *doubtful problems*, and uncertain ghesses [sic], and not as unquestionable conclusions, or matters of unconfutable science ....[31]

Another mid-century corpuscularian, Joseph Glanvill, puts the point similarly in his *Scepsis Scientifica* (1661). True philosophers, he says,

seek truth in the great book of nature, and in that search ... proceed with wariness and circumspection without too much forwardness in establishing maxims and positive doctrines ... [They] propose their opinions as hypotheses, that may probably be true accounts, without peremptorily affirming that they are.[32]

Like Locke after him, Glanvill insists that all scientific principles are conjectural; there is nothing we can see with certainty about the physical world:

For the best principles, excepting divine and mathematical [precisely Locke's exceptions], are but hypotheses, within which, we may conclude many things with security from error. But yet the greatest certainty, advanced from supposal, is still hypothetical. So that we may affirm that things are thus and thus, according to the principles we have espoused: But we strangely forgot ourselves, when we please a necessity of their being so in nature, and an impossibility of their being otherwise.[33]

All three writers – Boyle, Hooke, and Glanvill – were widely read in the 1660s and 1670s, when the ideas for the *Essay* were taking shape. It is likely that Locke knew the works of Hooke and Glanvill and it is certain that he knew Boyle's works, as the two were close friends for more than thirty years and, as Leyden has noted, Locke "followed with interest each new publication of his friend Robert Boyle."[34] Furthermore, Locke met frequently with Boyle's scientific circle at Oxford (of which Hooke was also a member) in the 1660s, and one presumes that among the topics for conversation was the nature of scientific knowledge and the tentative character of its hypotheses. It is not unreasonable to suggest that Locke's ideas on this topic stemmed, in part, from his discussions with, and readings of, the Oxford corpuscularians who, like him, were alarmed that hypothetical systems were being passed off as infallibly true theories.

(2) We have seen that, in explaining why the corpuscular philosophy can never be more than an hypothesis, Locke metaphorically likened nature to a clock whose internal mechanisms could never be observed. Just as we can only conjecture about the possible internal mechanisms of an unfamiliar clock, the scientist can only hypothesize about nature's hidden mechanisms. We know no more about the real natural processes than a 17th-century country bumpkin knew about the "famous clock at Strasbourg," which was accompanied by ingenious automata of every description. As we have seen, this clock-analogy[35] was widely exploited among Locke's predecessors – Descartes, Boyle, and Glanvill – in justification of their insistence on the necessarily hypothetical character of scientific principles.[36] Given the fact that this analogy occurred prominently in the works of several of the most important of Locke's predecessors, it is implausible that he did not borrow it from one of them.

(3) Locke was adamant in his insistence that hypotheses about submicroscopic events must construe atoms and their properties as natural extensions of the properties of macroscopic bodies. Locke believed there was a continuity in nature such that the laws governing macroscopic phenomena must be similar to, if not identical with, the laws governing submicroscopic phenomena. One finds elaborations of this argument in Descartes, Boyle, Hooke, and Newton.

It is well known that Descartes' scientific treatises are packed with analogies: the mind as a piece of wax, light transmission as a moving stick, light corpuscles as grapes in a vat. It is perhaps not so well known that analogy plays an important role in Descartes' philosophy of science as well. In the *Regulae*, for example, Descartes argued that since light is a process whose nature we cannot perceive directly, we should imagine light on the analogy of other natural agents which we do understand.[37] Again, in the *Dioptrique*, Descartes talks of the importance of analogies in understanding light.[38] But his most explicit discussion of analogies occurs in the *Principles*, which Locke undoubtedly read. There he writes:

> Nor do I doubt that anyone who used his reason will deny that we do much better to judge of what takes place in small bodies which their minuteness alone prevents us from perceiving, by what we see occurring in those that we do perceive than, in order to explain certain given things, to invent all sorts of novelties that have no relation to those that we do perceive.[39]

The sentiments of this passage are Locke's as well as Descartes'.

Boyle was the next to take up this theme. He argues that it is absurd to assume anything but that submicroscopic corpuscles obey the same laws as macroscopic bodies do. To say, for example, that the principles of mechanics apply to visible masses but not to invisible ones would be "as if a man should allow, that the laws of mechanism may take place in a town clock, but cannot in a pocket-watch".[40] For corpuscular hypotheses to be even intelligible, they must attribute the same type of behavior to atoms as we observe taking place in perceptible bodies.

In his *General Scheme* (c. 1667), Hooke too argues for the importance of models and analogies in the construction of hypotheses:

A most general help of discovery in all types of philosophical [i.e., scientific] inquiry is, to attempt to compare the working of nature in that particular that is under examination, to as many various, mechanical and intelligible ways of operation as the mind is furnisht with.[41]

Another of Locke's contemporaries who attaches importance to analogy in the argument from the seen to the unseen is Newton. The third of Newton's grand methodological precepts, the *Regulae Philosophandi*, runs as follows:

The qualities of bodies which admit neither intension nor remission of degree, and which are found to belong to all bodies within the reach of our experiments, are to be esteemed the universal qualities of all bodies whatsoever . . . nor are we to recede from the analogy of Nature, which uses to be simple, and always consonant with itself.[42]

However, regardless of whether Locke derived his beliefs about analogies from Descartes, Hooke, Boyle, or Newton, it is clear that the use of analogical hypotheses was a basic and explicit tenet of the corpuscularism which Locke so warmly embraced.

(4) Finally, I want to consider the likely sources of Locke's belief that corpuscular hypotheses must be compatible with "truths that are established in our minds" and with "other parts of our knowledge and observation".[43] The writer who comes most quickly to mind in this connection is Robert Boyle, who frequently emphasized that the two requirements of any sound hypothesis are that it accord both with other known laws and with observations. Boyle goes so far as to say that the function of an hypothesis is "to render an intelligible account of the causes of the effects, or phenomena proposed, without crossing the laws of nature, or other phenomena".[44] Elsewhere, he insists that a good hypothesis must "not be inconsistent with any other truth or phenomenon of nature".[45] Varying the wording yet again, he says that an hypothesis is acceptable if we can show "its fitness to solve the phenomena for which [it was] devised, without crossing any known

observation of law of nature".[46] Locke's remarks in the *Essay* at iv. 16. 12 seem to be little more than stylistic variations on this Boyleian theme.

The analysis offered here constitutes only a very partial account of Locke's views on methodological issues. But it should still be sufficient to make plausible the claim that Locke adhered to certain conventions of the philosophical and scientific *milieu* in which he matured. There was a cluster of beliefs – methodological as well as scientific – which many adherents of the "scientific philosophy" accepted; among them was the conviction that scientific knowledge was conjectural. In sharing this conviction, Locke was in distinguished company.

NOTES

1 In a classic piece of understatement, Locke speaks of himself as an "underlabourer" to the scientists Boyle, Sydenham, Huygens and "the incomparable Mr Newton" (*Essay*, ed. A. C. Fraser [Oxford, 1894], I, 14.).

2 Among those who have taken the above interpretation of Locke, the most prominent are probably R. I. Aaron, *John Locke* (Oxford, 1955), J. Gibson, *Locke's Theory of Knowledge* (Cambridge, 1960), and J. W. Yolton, *John Locke and the Way of Ideas* (Oxford, 1956).

3 Mandelbaum. *Philosophy, Science and Sense Perception* (Baltimore, 1964), pp. 1–60. I can do no better than cite Mandelbaum's own summary of his thesis: "The conclusion which I wish to draw . . . is that Locke, throughout his career, was an atomist, and that he accepted both the truth and the scientific usefulness (or, at least, the scientific promise) of the corpuscular, or new experimental, philosophy" (Ibid., p. 14).

4 R. M. Yost, Jr., 'Locke's Rejection of Hypotheses about Sub-Microscopic Events', *JHI* 12 (1951), 111–30.

5 Ibid., 111.

6 *Essay concerning Human Understanding*, iv, 3, para. 29.

7 Ibid., iv, 3, para. 9. Cf. also iv, 12, para. 10.

8 Ibid., iv, 3, para. 16.

9 Ibid., iv, 3, para. 25. (Italics added.) Elsewhere, he writes: "Thus, having no ideas of the particular mechanical affections of the minute parts of bodies, that are within our reach and view, we are ignorant of the constitutions, powers, and operations . . . "Ibid., iv, 3, para. 15.

10 As one of the numerous passages where Locke overtly takes a corpuscular view, consider his remark that heat and cold are "nothing but the increase or diminution of the motion of the minute parts of our bodies, caused by the *corpuscles* of any other body" Ibid., ii, 8, para. 21 (Italics added). Cf. also his *Elements of Natural Philosophy*: "By the figure, bulk, texture, and motion of these small and insensible corpuscles, all the phenomena of bodies may be explained."

11 "Therefore I am apt to doubt that how far soever human industry may advance useful and experimental philosophy in physical things, scientifical [knowledge] will still be out of our reach . . . " Ibid., iv, 3, para. 26.

[12] Locke expresses himself thus: "The understanding faculties being given to man, not barely for speculation but also for the conduct of his life, man would be at a great loss if he had nothing to direct him but what has the certainty of true *knowledge*. For that being very short and scanty, as we have seen, he would be often utterly in the dark, and in most of the actions of his life, perfectly at a stand, had he nothing to guide him in the absence of clear and certain knowledge. He that will not eat, till he has the demonstration that it will nourish him; he that will not stir, till he infallibly knows the business he goes about will succeed, will have but little else to do, but to sit still and perish" (Ibid., iv, 14, para. 11). Again, he remarks: "How vain, I say, it is to expect demonstration and certainty in things not capable of it, and refuse assent to very rational propositions . . . because they cannot be made out so evident, as to surmount even the least . . . pretence of doubting" (Ibid., iv, 11, para. 10).

[13] Ibid., iv, 14, para. 3.

[14] Ibid., iv, 14, para. 4.

[15] " . . . the propositions we receive upon inducements of *probability* are of *two sorts*, either concerning some particular existence, or, as it is usually termed, matter of fact, which, falling under observation, is capable of human testimony, or else concerning things, which being beyond the discovery of our senses, are not capable of any such testimony" (Ibid., iv, 16, para. 5).

[16] Cf. *Essay*, iv, 16, para. 2.

[17] " . . . effects we see and know; but the causes that operate and the manner they are produced in, we can only guess and probably conjecture" (Ibid., iv, 16, para. 12).

[18] Ibid. Elsewhere he writes that an accomplished experimenter can often make valuable hypotheses: "I deny not but [that] a man, accustomed to rational and regular experiments, shall be able to see further into the nature of bodies, and guess righter at their yet unknown properties, than one that is a stranger to them; but yet, as I have said, this is but judgment and opinion, not knowledge and certainty . . . [hence] natural philosophy is not capable of being made a science" (Ibid., iv, 12, para. 10). Again, he notes: "Possibly inquisitive and observing men may by strength of judgment, penetrate further, and, on probabilities taken from wary observation, and hints well laid together, often guess right at what experience has not yet discovered to them. But this is but guessing still; it amounts only to opinion, and has not that certainty which is requisite to knowledge" (Ibid., iv, 6 para. 13).

[19] Yost, op. cit., 127.

[20] Ibid., 125. Further on, Yost puts it this way: "Speaking generally, Locke said that the observable clues are so scanty that there is no hope of making good guesses about any kind of specific submicroscopic mechanisms" (Ibid., 126).

[21] *Essay*, iv, 12, para. 13. He prefaces the quoted passage by an endorsement of hypotheses which is quite unequivocal: "Not that we may not, to explain any phenomena of nature, make use of any probable hypothesis whatsoever" (Ibid.).

[22] Yost, op. cit., 127.

[23] *Essay*, iii, 6, para. 9. Cf. also iii, 6, para. 39.

[24] Ibid., iv, 3, para. 25.

[25] "And at least that we take care that the name of *principles* deceive us not, nor impose on us, by making us receive that for an unquestionable truth, which is really at best but a doubtful conjecture; such as are most (I had almost said all) of the hypotheses in natural philosophy" (Ibid., iv, 12, para. 13).

26 Cf. Ibid., iv, 20, para. 11.
27 Ibid., iv, 12, para. 13.
28 For some of the relevant passages in which Locke takes a corpuscular position, cf. *Essay* ii, 8, paras. 13–21 and ii, 21, para. 75; iii, 6, para. 6; iv, 13, para. 16 and para. 25; iv, 10, para. 10, iv, 16, para. 12; iv. 6, para. 10 and para. 14. As Locke's 19th-century editor, Fraser, notes: "It is to the 'corpuscularian hypothesis' that he [Locke] appeals in the many passages in the *Essay* which deal with . . . the *ultimate physical cause* of the secondary qualities . . . " (*Essay*, (ed. Fraser), vol. ii, 205 n.).
29 *Essay*, iv, 12, para. 13.
30 Robert Boyle, *Works*, ed. Birch (London, 1772), I, 303.
31 R. Hooke, *Micrographia* (London, 1667), preface, n. p.
32 *Scepsis Scientifica* (London, 1665), 44.
33 Ibid., 170–1.
34 J. Locke, *Essays on the Laws of Nature* (ed. Leyden, Oxford, 1954), 20. Maurice Cranston, Locke's biographer, records that "Locke, as Boyle's pupil, absorbed much of the Boylian conception of nature before he read Descartes and became interested in pure philosophy." *John Locke* (London, 1957), pp. 75–76.
35 Locke uses the analogy on at least two occasions: Essay, iii, 6, para. 9; iv, 3, para. 25.
36 See the preceding chapter for the relevant texts.
37 "If he [the scientist] finds himself . . . unable to perceive the nature of light, he will, in accordance with Rule 7, enumerate all the natural forces in order that he may understand what light is, learning its nature, if not otherwise, at least by *analogy* . . . from his knowledge of one of the other forces" (*Oeuvres* (ed. Adam & Tannery), X, 395. Italics added).
38 He says that he intends to offer "two or three comparisons, which will help us to understand it [viz., light] in the most convenient manner, in order to explain all those of its properties that experience allows us to know, and to deduce thereafter all the others which may not be so easily noticed . . . ." (Ibid., VI, 83).
39 *Principles*, Part iv, para. 201.
40 *Works*, IV, 72.
41 Hooke, *Posthumous Works*, ed. Waller (London, 1705), 61. Italics in original.
42 I. Newton, *Principia Mathematica* (trans. Motte), 385). It should be pointed out that in the first edition of the *Principia* (1687), which was the only one to appear in Locke's lifetime, this principle did not appear. Newton introduced it in the 1713 edition of the *Principia.*
43 *Essay*, iv, 16, para. 12.
44 *Works*, iv, 234.
45 Ibid., i, 241.
46 Ibid., iv, 77.

CHAPTER 6

HUME (AND HACKING) ON INDUCTION

INTRODUCTION

Several years ago, Ian Hacking wrote a fascinating book called *The Emergence of Probability* (1975). It deals with several important issues in the history of epistemology and breaks new ground in its treatment of most of them. But in one crucial respect, Hacking's analysis restates a certain pervasive philosophical myth. More than restating the myth, Hacking gives it as forceful an expression as one finds in the literature. That myth, closely related to the purist model I described in Chapter 2, concerns David Hume and the so-called problem of induction.

According to this myth, there is a univocal problem of induction, no one before Hume realized there to be such a problem, and Hume 'discovered' it in the late 1730s. Hacking formulates this 'myth' in the form of two historical theses:

(1) "Until the 17th century [at one point he dates it 'around 1660'[1]], there was no concept of evidence with which to pose the problem of induction!"[2]

(2) "The sceptical problem . . . of induction, was first published in 1739, in David Hume's *A Treatise of Human Nature*."[3]

It seems to me that both theses, although commanding broad assent among contemporary philosophers, are mistaken. Worse than that, they are wildly mistaken, for taken together they rest on three different but equally significant confusions.[4] To begin with, by alluding to an omnibus 'problem of induction', they paper over what are really a wide variety of epistemic issues. Secondly, they ignore the long pre-Humean preoccupation with the problem of theory validation. Thirdly, these theses give to Hume's discussion of induction an importance and significance which it scarcely deserves historically.

However, before I can deal specifically with Hacking's two theses, some elementary but important distinctions need to be drawn. (Indeed, a failure to draw them clearly has led Hacking and many of his predecessors to conflate several historically distinct lines of development, which ought to be clearly distinguished.)

TWO PROBLEMS OF INDUCTION

It has been fashionable since Peirce and Keynes to construe 'induction' so broadly as to include virtually all forms of non-deductive inference. Doubtless useful for certain purposes, this subsumption of all manner of sins under one category creates certain difficulties. In particular, it disposes us to imagine that 'the problem of induction' has a univocal sense, when in fact there are probably as many problems of induction as there are species of ampliative inference. Amongst all the different problems of induction, there are two which deserve, and have attracted, special attention. The first of these is Hume's; for reasons that will become clear later, I shall refer to this as the 'plebian problem of induction'. It can be formulated in the following way:

Plebian Problem of Induction	Given a universal empirical generalization and a certain number of positive instances of it, to what degree do the latter constitute evidence for the warranted assertion of the former?

The positive instances of a plebian induction constitute partial, but *direct* evidence for the association of the properties conjoined in the generalization.

There is, however, a second type of inductive or ampliative inference, which we might call '*induction to theories*' or '*aristocratic induction*'. In such cases, we are testing a theoretical statement, i.e., a statement containing some terms which have no direct linkages, via correspondence rules, with observable properties. Theoretical statements of this kind, even though they lack direct observational analogues, can often be utilized to make predictions about obsrvable processes.

Consider for instance, a pair of examples that were well known in Hume's time:

(1) In *Principia*, Newton explained what we call Boyle's law by developing a model which conceives of gases as composed of imperceptible particles which repel one another with a force inversely related to the distance between them. Newton shows that a gas, so conceived, would indeed obey the familiar $pV = k$ law.

(2) René Descartes sought to explain planetary astronomy by assuming that all the planets are carried around the sun by a vortex, i.e., a rotating fluid composed of imperceptible particles. Since all the particles of this vortex were said to move in the same direction, it followed that all the large bodies carried by this medium would move in the same direction and in approximately the same plane of revolution. The observable fact that the

planets behave in this way was taken to be confirmatory of Descartes' hypotheses about the microstructure of the spatial region of the solar system.

In both of these cases, as well as dozens of others that could be cited, the evidence for a theoretical statement is *indirect* rather than direct. The evidence establishes that one entailment of the statement is true, but that true entailment does *not* count as positive instance of the statement, if we understand a positive instance of a universal of the form '$(x)(Ax > Bx)$' to be 'Aa & Ba'. (This is not to suggest, of course, that the statements which confirm theories are not positive instances of *some* statement or other. The statement that the doubling of the volume of a gas halved its pressure is a positive instance of Boyle's law; but it is (in the technical sense) *not* a positive instance of Newton's assumption that gaseous particles repel one another. Similarly, to remark that Mars and Jupiter move in the same direction about the Sun is to give a positive instance for the generalization that "all planets move in the same direction"; but it is *not* a positive instance for the vorticular hypothesis.)

The point of this belabored distinction is simply this: genuinely theoretical statements may have confirming instances (i.e., known, true empirical statements entailed by them), but they do not possess positive instances. Thus, when we talk about aristocratic induction to theories, we must formulate a problem different from the plebian one; specifically,

Aristocratic Problem Induction	Given a theory,[5] and a certain number of confirming instances of it, to what degree do the latter constitute evidence for the warranted assertion of the former?[6]

The philosophical issues posed by the problems of plebian and aristocratic induction are significantly different. This becomes clear if we consider what would count as solutions to the two problems. For instance, if we could establish in an appropriate sense of the term, that nature was indeed 'uniform', then we would have a solution to the plebian problem. But no amount of nature's uniformities will solve the aristocratic problem, because what is at stake there is not whether the world will continue to be the way it has been in observed cases; rather, with aristocratic induction, we do not know – even in those cases where we have a confirming instance – that all the assertions made by the theoretical statement about a given state of affairs were true in the observed cases. Concerning any observable state of affairs, a theoretical statement makes two sorts of claims: a directly testable one about observable relations *and* a claim (not directly testable) about microstructural or otherwise inaccessible processes. Where with plebian induction, we know that if all the possible positive instances of a general statement are true, then the statement itself is true (since it asserts nothing more than the sum of its possible positive instances), it is entirely conceivable that *all* the possible confirming instances of a theoretical statement could be true and yet the statement could still be false (since theoretical statements typically assert more than the conjunction of their possible confirming instances).

Such differences as these between plebeian and aristocratic induction are as significant historically as they are philosophically, for the developmental lines that led to the emergence of plebeian induction are quite distinct from those that led to aristocratic induction. For instance, virtually all of Hume's paradigm cases of induction involve plebeian induction. Thus, when he asks, how do I know that this bread which nourished me today will nourish me tomorrow, or how do I know that fire, which has burned flesh in the past, will burn me next time, he is envisaging a situation in which we have a certain number of (unproblematic and unambiguous) positive instances of the conjunction of two observable properties. Within that context the plebeian problem is: what credence do those positive instances lend to the universal generalization whose antecedent and consequent they instantiate? Plebeian induction, so conceived, *is a feature of only (but not all) those ampliative inferences which deal exclusively with observable events, objects or processes.*[7]

Now this limitation on plebeian induction is a little embarrassing for Hume, or at least it ought to have been, since most of the best known theories of Hume's day – including those of Newton, Descartes, Boerhaave, Huygens, and Boyle – did not consist primarily of statements which could be said to have positive instances. These theories involved numerous statements about various unobservable entities – atoms, subtle fluids, imperceptible forces, and the like.

Ian Hacking's analysis, while purporting to tell the early history of induction, focusses exclusively on plebeian induction and ignores aristocratic induction altogether. Apparently persuaded by Hume that plebeian induction is the primary form of inductive inference, Hacking limits his story to looking for Hume's predecessors. Because there are few, Hacking can get much mileage out of the two theses enumerated above. What handicaps his analysis is his failure to consider the history of aristocratic induction; for, had he explored its history and pre-history, he would have discovered that (a) *there was a concept of (aristocratic) inductive evidence long before the 1600s*, and (b) *there was a sceptical problem of (aristocratic) induction long before Hume*.

Was there a concept of (inductive) evidence prior to 1660? Hacking's claims on this issue are quite startling. "A concept of evidence", he tells us, "is a necessary condition for the stating of a problem of induction. A problem of induction does not occur in the earlier annals of philosophy [i.e., long before Hume] because there was no concept of evidence available".[8] In asserting that there was no concept of evidence before the Port-Royal *Logic*, Hacking is not denying that people were prepared to defend their beliefs by citing testimony and authority, both of which were regarded as rendering beliefs authentic or 'probable'. What was lacking, in Hacking's view, is that conception of evidence which "consists in one thing pointing beyond itself".[9] Specifically, what he sees as lacking was the view that one thing (or statement) could be taken as evidence for another. Is smoke evidence for fire, is thunder evidence for lightning? These and a host of similar questions received, on Hacking's reading of the history, neither answer nor even coherent formulation until the later 17th century.

This claim is more than a little remarkable. On purely *a priori* grounds, one wants to say it must have been otherwise. Intelligent men cannot think about the nature of knowledge for two millenia without coming to grips with such questions. But history has a habit of confounding our intuitions and if Hacking's claim is to be faulted, the case must rest upon more than hunches about human nature. One's first inclination is perhaps to think of Aristotle, the Stoics and the Galenists, all of whom had much to say about evidential 'signs' and confirming instances. After all, Sextus Empiricus had argued at length – as had the Greek medical tradition before him – that in the process of diagnosis, the physician (and by implication every natural scientist) takes one state (e.g., fever) to be evidence for another. Hacking brushes aside such counter-examples by saying that he is making "no claim about Sanskrit or Greek concepts of evidence. I am concerned with a specific lack at a particular time .... "[10] If we could assume that the medieval and Renaissance philosophers and scientists (who, according to Hacking, lacked a concept of evidence) were completely disconnected from their Greek heritage, this might be a sensible suggestion. But such an assumption is palpably false. All through the later Middle Ages, the writings of Avicenna and Averroes were widely known throughout western Europe, including those of their works which alluded to Galenic theories of inference and evidence. As for the specific tradition of ancient scepticism, it has been unambiguously established that it came into the West well before Hacking acknowledges its presence; indeed, the appropriate writings of Sextus Empiricus were widely available from the middle of the 16th century.[11] Hence Hacking's dismissal of the ancient tradition as irrelevant to an understanding of early modern discussions of the philosophy of science will not do.

But even if we were to ignore the sceptical tradition, Hacking's claim that a concept of inductive evidence was absent until the middle of the 17th century seems impossible to square with a number of well-known medieval, renaissance and early 17th-century texts, all of which utilize a concept of empirical evidence of precisely the kind whose presence Hacking denies. To locate this concept of evidence, however, we must go back to our earlier distinction between plebeian and aristocratic induction. Without saying so, Hacking has focused almost exclusively on the former of the two; that is, he has looked for writers who see evidence in terms of the relation between an empirical universal generalization and its positive instances. Given this constriction of the issues, his historical claims are almost right. There were relatively few thinkers before 1660 (Bacon would be an exception)[12] who worried about the degree to which positive instances probabilified generalizations. (This is also hardly surprising since, unlike the plebeian Hume, they believed science to consist of grander things than empirical generalizations.) But there were a great many thinkers who worried about evidence in what I have been calling the aristocratic sense. Indeed, *one of the core problems in medieval, renaissance and early 17th-century epistemology was precisely this: to what degree, if at all, does a confirming instance of a theory contribute to the cognitive well-foundedness of that theory*?

To see how and why this is so, the best place to start is probably with Locke's *Essay*. Although written at the end of the period in question, it provides a convenient set of distinctions to utilize for describing and unpacking the earlier material.

In that section of the *Essay* concerned with knowledge and opinion, Locke speaks of the 'degrees of assent' appropriate to different forms of belief. He distinguishes, in matters of probability, two central domains: conjectures about ascertainable matters of fact and speculations concerning things "beyond the discovery of our senses".[13] The first category, of course, corresponds to plebeian induction: that swallows come in the summer, that fire warms flesh, that iron sinks in water. Such general statements about the observable properties of sensible bodies represent, for Locke, one species of probable judgments. But Locke insists that such cases are by no means the only type in which questions of probability arise. We are also dealing with probable opinions whenever we seek to explain the causes of any observable process in terms of entities "which sense cannot discover".[14] Locke's examples of this sort of probability make it clear that he is speaking primarily of the scientific explanations of his day: the explanation of heat on the assumption that it "consists in a violent agitation of the imperceptible minute parts of the burning matter",[15] the nature of liquefaction, animal embryology; these and other questions "concerning the manner of operation in most parts of nature"[16] are open only to "guess" and "conjecture". Nonetheless, Locke wants to insist that theories of this kind do, under appropriate circumstances, acquire a certain probability and "this sort of probability, which is the best conduct of rational experiments, and the rise of hypotheses, has also its use and influence".[17]

What Locke is saying in the *Essay* is that there are really two problems of induction: one, plebeian induction, having to do with assessing the probability of statements about observable matters of fact; the second and scientifically more important type, having to do with assessing the probability of theoretical statements (in the genuine sense of 'theory'). Now, Hume and Hacking really consider only the first to be the problem of induction, whereas, in fact, most of the early epistemologists of science reckoned that the second problem was of greater philosophical moment. And it is with respect to the latter problem that we can begin to see what kinds of inductive evidence and what kinds of inductive scepticism formed the backdrop to Hume's re-definition and constriction of the problem of induction.

As we have already seen, the structure of aristocratic induction involves us in attempting to draw conclusions about the truth status of a theoretical statement on the strength of its known confirming (but not positive) instances. There are two major difficulties here, ones which have been familiar to scientists and philosophers since antiquity. The first of these difficulties is, of course, that true conclusions can be drawn from false premises. Hence, it is easy to see that a true confirming instance is entirely compatible with a false theory and that *no* number of confirming instances can establish the truth of a theory. But mightn't we still be able to say that a sufficiently large number of confirming instances does at least make it likely that a theory is true?

There were many thinkers who subscribed to just such a point of view, i.e., who maintained that a sufficiently large number of confirming instances constituted evidential grounds for the warranted assertion of a theory. In a discussion that was well known throughout the late Middle Ages, Galen insisted that the proper method for science was to utilize one's knowledge of existing and observable signs or effects in order to uncover theories about the covert and imperceptible causes of these effects.[18] Averroes and his followers in the Middle Ages and Renaissance made much of the method of "demonstration by signs", by which they meant probable arguments from observables to unobservables.[19] Robert Grosseteste had argued that we can come to a 'probable knowledge' of causes by carefully studying their effects.[20]

By the Renaissance and throughout the 17th century, this argument took a more specific form. In particular, many thinkers argued that although a false theory might have some true consequences, it would be highly unlikely – in the case of a false theory – that after a large number of its consequences were examined, they would all turn out to be true. Put differently, it was commonly claimed that a theory that had many known true consequences and no (known) false ones could be asserted, with high confidence (the usual language here is "with moral certainty") to be true. Consider, for instance, the well-known text from the Renaissance astronomer Christopher Clavius:

> If [Copernicus' critics] cannot show us some better way [to save the phenomena], they certainly ought to accept this way, inferred as it is from so wide a variety of phenomena . . . if one is not justified in inferring on the basis of the phenomena that there are eccentrics and epicycles in the heavens, just because a true conclusions may be inferred from a false premise, then the whole of natural philosophy will be destroyed. For in precisely the same manner, when anyone infers from a known effect that this is not true, because a true conclusion can be inferred from a false premise, and thus all the principles of natural science that have been discovered by philosophers will be destroyed.[21]

Kepler's view was similar. After insisting that we should "attempt to confirm and establish [hypotheses] by means of the appearances,"[22] he argues that the Copernican system has been empirically established in just such a manner. Kepler's contemporary Francis Bacon insisted, in the *Novum Organum*, that a theory must always be tested against data that were not utilized in its construction. Specifically, for any theory, "we must observe whether, by indicating to us new particulars, it confirms that wideness and largeness by a collateral security".[23]

René Descartes, whom Hacking excludes from his story because he "has no truck with the nascent concept of probability",[24] makes a similar point concerning the capacity of a large class of confirming instances to dispose us to accept the theory they confirm as legitimate:

> Although there exist several individual effects to which it is easy to adjust diverse causes [i.e., hypotheses], one to each, it is however not so easy to adjust one and the same [hypothesis] to several different effects, unless it be the true one from which they proceed.[25]

It was Boyle, too, who had specified that the test of any "excellent hypothesis" involved determining whether it could successfully "foretell future phenomena".[29] Boyle's colleague, Robert Hooke, likewise stressed that although scientific knowledge is uncertain, theories can accumulate confirming instances which can render them very probable:

> . . . the Nature, Composition and internal operations and Powers of mixt Bodies are far beyond the reach of the Senses . . . And upon that account the *Data*, upon which the Ratiocination is founded, being uncertain and only conjectural, the Conclusions or Deductions therefrom can at best be no other than probable; but still *they become more and more probable, as the Consequences deduc'd from them appear . . . to be confirm'd by fact or Effect.*[30]

Further texts illustrative of these same themes could be cited at much greater length, and several of them appear in earlier chapters. But even this brief collection of texts should establish that, long before Hacking's mysterious date of 1660, there were numerous scientists and philosophers who certainly possessed a notion of empirical evidence, i.e., who subscribed to the view that a large number of confirming instances for a theory might reasonably be taken as a ground for believing or accepting the theory in question. Indeed, it is straight out of this centuries-old tradition that Locke's discussion of the probability of hypotheses emerged.

But this is only half the story. That there was a centuries-old concept of evidence (not that different in its essentials from our own), I take as established. But that still leaves us with the familiar claim, reiterated by Hacking, to the effect that it was Hume who discovered the "sceptical problem of induction". It is to that issue that I now turn. Associated with both plebeian and aristocratic methods of induction are characteristic sceptical problems. The plebeian sceptic asserts that the positive instances of an inductive generalization provide no grounds whatever for asserting the truth or likelihood of the generalization; the aristocratic sceptic asserts that confirming instances of a theory provide no grounds for asserting its truth or likelihood. That these sceptical challenges are different is undeniable; but that they both constitute challenges to inductive or ampliative inference should be clear. What I want to show in this section is simply that *long before Hume, scepticism about aristocratic induction was rampant and that detailed cogent arguments, every bit as compelling as Hume's critique of plebeian induction, had been adduced to show that aristocratic induction could not produce knowledge or even reliable opinion.*

In its most general form, the sceptical challenge to aristocratic induction can be formulated as follows:

> Given any theory, T, and any body of evidence, E, which T entails, there are (possibly infinitely) many other theories,

contraries of T, which will also entail E. Hence, no amount of evidence can render T likely or probable.

There were two closely related considerations that led many thinkers to this sceptical conclusion. The first was a general recognition that false theories could have true consequences. From at least the time of Aristotle, every schoolboy knew that a statement with true entailments might well be false. This, in its turn, was taken by many thinkers to mean that the possession of true confirming instances by a theory was no definitive sign of its truth. But the fallacy of affirming the consequent will take us only so far towards the sceptical handling of aristocratic induction. After all, one might grant that confirming instances do not prove a theory to be true and still assert that a large number of such instances constitutes ground for asserting the theory to be likely (as we have seen above, many thinkers were inclined to make just such an assumption). The really decisive consideration which historically led to aristocratic scepticism was *the emergence*, first within astronomy and later in the other theoretical sciences, *of viable contrary theories which were presumed to be observationally equivalent*, with respect to any possible body of evidence. The story probably begins with Apollonius, who demonstrated that the motion of a planet on an epicycle and its motion on an eccentric (of appropriate magnitude) entailed precisely the same observational consequences. In short, experience, no matter how diverse or extensive, no matter how shrewdly collected, could not conceivably discriminate between the likelihood or verisimilitude of an eccentric or an epicyclic hypothesis, and could never make either hypothesis more likely than not.

The incapacity of experience to adjudicate between different astronomical hypotheses about the motion of heavenly bodies was much discussed in the Middle Ages and the Renaissance. For instance, in his commentary on *De Caelo*, Aquinas wrote:

the assumptions of the astronomers are not necessarily true. Although these hypotheses appear to save the phenomena, one ought not affirm that they are true, for one might be able to explain the apparent motions of the stars in some other way . . .

In *Summa Theologica*, he made the same point:

[one] way of accounting for a thing consists, not in demonstrating its principle by a sufficient proof, but in showing which effects agree with a principle laid down beforehand. Thus, in astronomy we account for eccentrics and epicycles by the fact that we can save the sensible appearances of the heavenly motions by this hypothesis. But this is not really a probative reason, since the apparent motions can, perhaps, be saved by means of some other hypothesis.[32]

The Renaissance Averroist, Agostino Nifo, points out the epistemic dilemma of the theoretician in a slightly different way:

> Those therefore err who, starting out from a proposition [i.e., a statement of evidence] whose truth may be the outcome of various causes, decide definitely in favor of one of these causes. The appearances can be saved by the sort of hypotheses we have been talking about, but they may also be saved by other suppositions not yet invented.[33]

What both Aquinas and Nifo are arguing is that for any theory, *T*, and any set of its confirming instances, *E*, it is possible that there are indefinitely many contraries of *T*, all of which, like *T*, entail *E*. Under those circumstances, there is no warrant for taking *E* as indicating either the truth or the likelihood of *T*.

This sceptical approach to aristocratic induction originally centered on astronomical theories; to a degree, it was at the core of the instrumentalist-realist controversy in the history of astronomy whose story had been recounted in such detail by Pierre Duhem.[34] But it is important to understand that by the 16th and 17th centuries, it had moved well out of astronomy into a general critique of the role of empirical evidence in authenticating theoretical claims, e.g., at the hands of Cusa and Coronel.

This became especially common by the first half of the 17th century, when the emergence of the mechanical philosophy brought theorizing about unobservable entities to a fine art. Whether one looks to pneumatics, optics, mechanics, or chemistry, the story was much the same by the 1630s and 1640s: in one domain of science after another, assumptions were being made about the fine structure of nature in order to explain observable effects. Specific hypotheses about the sizes, shapes, velocities, and other properties of these micro-entities, could only be very indirectly 'tested' by probing their observable consequences. But few natural philosophers felt they could convincingly argue that, for any given assumption about elemental attributes, there is not an unspecifiably large number of equiprobable, contrary assumptions, all of which would be equally compatible with the available evidence. Under such circumstances, it was tempting to take the sceptical position that there was no positive evidence that could make a decisive difference. Such views are neatly summarized towards the end of Descartes' *Principles* (1645), where he writes:

> it may be said that although I have shown how all natural things may be formed, we have no right to conclude on this account that they were produced by these causes . . . This I most freely admit; and I believe that I have done all that is required of me if the causes I

have assigned are such that they correspond to all the phenomena manifested by nature without inquiring whether it is by their means or by others that they are produced.[35]

What Descartes is conceding here, as he does elsewhere, is the fact that "there is an infinity of different ways in which all things that we see could be formed".[36] This Cartesian concession amounts to the claim that there are indefinitely many contrary hypotheses which are confirmed by equal numbers of instances. Though not happy with this result, Descartes' text acknowledges what could be convincingly shown if more space were available; namely, that sceptical arguments against the efficacy of (aristocratic) induction were the stock in trade of almost every 17th-century philosopher. (Much of the relevant evidence is presented elsewhere in this book, especially in Chapters 4, 5 and 8).

Contrary to Hacking and the orthodox history of philosophy, Hume was scarcely needed to draw the attention of scientists and philosophers to the fact that the issue of the evidential support for scientific theories was a pressing and an unresolved one.

Indeed, what Hume's work did – and continues to do – is to direct our attention away from the pressing problems of aristocratic induction and towards the comparatively trivial problems of plebeian induction.[37] Invoking the authority and the precedent of Hume, we have convinced ourselves that the chief worry about induction is the one of temporal extrapolation of observed conjunctions. Hume's predecessors, far more than many of his successors, saw that the relations between theory and evidence pose much weightier problems than ever occurred to that dour scientific dilettante from Edinburgh.

Hume's avoidance of aristocratic induction was probably related to his almost unparallelled ignorance of the science of his time.[38] But it is due far more to his curious sensationalist epistemology, entailing as it does that genuine theories (i.e., those involving theoretical entities) are improper. For Hume, there is no worry about 'induction to theories', because his account of knowledge leaves no room for such theories. What is remarkable about all this is that generations of philosophers of science – many of whom find Hume's sensationalism repugnant – have nonetheless accepted Hume's insistence that plebeian induction is archetypal.[39] We have rejected Hume's plebeian epistemology, in large measure because it offers an impoverished account of scientific knowledge; yet we have retained his formulation of the inductive problem, refusing to face up to the equal injustice it does to an understanding of scientific inference.

NOTES

[1] Hacking (1975), p. 9.
[2] Ibid., p. 31.
[3] Ibid., p. 176.
[4] This myth about "Hume's problem" is closely linked both to issues of Galilean mechanics and to the logic of discovery which I discuss in Chapters 3 and 11 respectively.
[5] It is important to stress that a 'theory' in this sense must postulate one or more unobservable entities, i.e., statements which *could* arise as empirical generalizations do *not* count as theories for these purposes.
[6] What I am calling 'aristocratic induction' is similar to what Mandelbaum (1964) has called 'transdiction'.
[7] As T. Seidenfeld has pointed out to me, there are many theories dealing with exclusively observable entities (e.g., statistical theories) which do not possess positive instances and thus equally fail to exemplify plebeian induction.
[8] Hacking (1975), p. 32.
[9] Ibid., p. 34.
[10] Ibid.
[11] Cf., for instance, Popkin (1968) and Schmitt (1972).
[12] Cf. Bacon (1857–59), I 25, 82, 103, 106 and 117.
[13] Locke (1894), iv. 16. para. 5.
[14] Ibid., iv. 16, para. 12.
[15] Ibid.
[16] Ibid.
[17] Ibid.
[18] Cf. Walzer (1944).
[19] Cf. Crombie (1953) and Randall (1961).
[20] Cf. Crombie (1953).
[21] Quoted in Blake, Ducasse and Madden (1960), p. 33.
[22] Ibid., p. 44.
[23] Bacon (1857–59), I, 106.
[24] Hacking (1975), p. 45.
[25] Descartes (1897–1957), II, p. 199.
[26] Ibid., IX, p. 123. Cf. Buchdahl (1969) and Chapter 4 above.
[27] Boyle (1772), IV, p. 234.
[28] Ibid.
[29] *Royal Society, Boyle Papers*, vol. IX, f. 25.
[30] Hooke (1705), pp. 536–637, my italics.
[31] Aquinas, quoted by Duhem (1969), p. 41.
[32] Ibid., p. 42.
[33] Ibid., p. 48.
[34] Cf. Duhem's (1969) and his (1913–1954), and Mittelstrass (1962).
[35] Descartes (1931), I, p. 300.
[36] Ibid.
[37] For an interesting exception, see Niiniluoto and Tuomela (1973).
[38] Indeed, it is difficult to find a major philosopher between Socrates and G. E. Moore who knew less than Hume about the science of his time!

[39] This way of looking at the matter is in sharp contrast to Stove's complaint that "Hume's discussion of our 'reasonings concerning matters of fact' never received, from the great 19th-century writers on scientific inference, the attention due it" (1968) p. 187.

REFERENCES

Bacon, F., *Works*, (ed. Spedding, Ellis and Heath), London, 1857–59, 14 vols.

Blake, R., Ducasse, C., and Madden, E., *Theories of Scientific Method: The Renaissance through the 19th Century*. Seattle, 1960.

Boyle, R., *Works* (ed. Birch), London, 1772, 4 vols.

Buchdahl, G., *Metaphysics and the Philosophy of Science*, Cambridge, Mass., 1969.

Crombie, A., *Robert Grosseteste and the Origins of Experimental Science*, Oxford, 1953.

Descartes, R., *Oeuvres* (ed. Adam and Tannery), Pris, 1897–1957.

Descartes, R., *Philosophical Works* (ed. Haldane and Ross), Cambridge, 1931.

Duhem, P., *Le Système du Monde*, Paris, 1954–59.

Duhem, P., *To Save the Phenomena*, Chicago, 1969.

Hacking, I., *The Emergence of Probability*, Cambridge, 1975.

Hooke, R., *Posthumous Works*, London, 1705.

Locke, J., *Essay Concerning Human Understanding* (ed. Fraser), Oxford, 1894, 2 vols.

Mandelbaum, M., *Philosophy, Science and Sense Perception*, Baltimore, 1964.

Mittelstrass, J., *Die Rettung der Phaenomene*, Berlin, 1962.

Niiniluoto, I. and Tuomela, R., *Theoretical Concepts and Hypothetico-Inductive Inference*, Dordrecht, 1973.

Popkin, R., *History of Scepticism from Erasmus to Descartes*, New York, 1968.

Randall, J., *The School of Padua*, Padova, 1966.

Schmitt, C., 'The Recovery and Assimilation of Ancient Scepticism in the Renaissance', *Rivista Critica di Storia della Filosofia*, 1972, 363–84.

Stove, D., 'Hume, Probability, and Induction', in V. Chappell (ed.), *Hume* (London, 1968), 187 ff.

Walzer, R., *Galen on Medical Experience*, Oxford University Press, 1944.

CHAPTER 7

THOMAS REID AND THE NEWTONIAN TURN OF BRITISH METHODOLOGICAL THOUGHT

INTRODUCTION

In a famous passage in the preface to his *Treatise*, Hume expressed the fervent hope that he could do for moral philosophy what Newton had done for natural philosophy.[1] In 18th-century ethics, literature, political theory, theology, and of course, natural science, similar sentiments were expressed openly and frequently.[2] Newton's *Principia* seemed to have established, almost overnight, new standards for rigor of thought, clarity of intuition, economy of expression and, *above all*, for the certainty of its conclusions. At long last, natural philosophy, which had hitherto been open to such controversy and speculation, was established on a secure foundation. It was tempting to believe that conjecture had given way to demonstration and that an infallible system, based on rigorous inductions from experimental evidence, had finally been devised.[3] Outside of the natural sciences, where Newton's real achievements were obscured by what scientific non-initiates took them to be, the enthusiasm for Newton reached an even higher pitch. Newton's great contribution, it was said, was not so much his cosmological synthesis *per se*, but rather the formulation of a new conception of science and its methods. Newton was seen as the harbinger of an inductive, experimental learning which proceeded by a gradual ascent from the particulars of observation to general laws which were true and virtually incorrigible. What Bacon had prophesied in the way of an inductive interpretation of nature, Newton had brought to fruition.

Not surprisingly, therefore, many of Newton's casual remarks on scientific method, especially the famous "hypotheses non fingo", were taken up as slogans and catch-phrases for intellectual reformers in almost every branch of human thought, especially in Britain. The 18th century has often been called the Age of Newton, and rightly so; but it was Newtonian, not so much in its physics or its metaphysics, as in its conception of the aims and methods of science. Putting it another way, it was Newton's inductivism and experimentalism – in short, his peculiar kind of empiricism – rather than his optics or his mechanics that motivated the leaders (and the charlatans) of 18th-century English intellectual history.

There is, of course, nothing particularly new or original about the brief account I have just given; to the contrary, it represents the view of most historians who have studied the period. But there is a curious flaw in this oft-repeated account of the development of British thought in the Enlightenment; namely, where do the professional philosophers come into the picture and how was the "Newtonian message" (which, as we have seen, was primarily a methodological and epistemological one) received among British philosophers? Although histories of philosophy often bracket Newton with the classical British empiricists, Locke, Berkeley, and Hume, such a conjunction is more misleading than illuminating, at least so far as the history of the philosophy of science is concerned. Indeed, those three empiricists are markedly un-Newtonian when it comes to questions of scientific method. Locke, for instance, died before most of Newton's pronouncements on methodology were published, so we look in vain for signs of Newtonian influence there.[4] Berkeley, on the other hand, though undoubtedly aware of Newton's inductive empiricism, developed a theory of scientific method and concept formation which is almost as alien to Newton's views as any could be.[5] Indeed, Berkeley's *De Motu* can be read as a thinly disguised critique of Newtonian empiricism and inductivism. The situation is not vastly different with Hume, who seems to have taken little or no cognizance of Newton's numerous methodological *obiter dicta*. In fact, when Hume did come to grips with methodological issues (e.g., induction or causality), his conclusions were diametrically opposed to the then usual interpretation of Newton's doctrines. Of course, I do not mean to suggest that Newton's physics, his theory of space and time, or his theology were without influence on Locke, Berkeley, and Hume; on the contrary, his work was profoundly important on each of these scores. But, and this is the anomaly, on the crucial questions of scientific method and the epistemology of science – the very areas where Newton ventured most openly and frequently into the epistemologist's domain – there seems to be little evidence indeed that the classic British empiricists were either very impressed by, or paid heed to, Newton's much publicized views.

Within the next century, however, circumstances were to change considerably. English philosophical works of the early and middle 19th century are teeming with references to Newton's philosophy of science, to his *Regulae Philosophandi*, to the 'General Scholium' to the *Principia*, and to Queries XXVIII and XXXI of the *Opticks* (where Newton discusses the methods of science). Indeed, almost all of the major figures of 19th-century British philosophy, logic, and philosophy of science (e.g., Brown, Herschel, Stewart,

Whewell, Mill, Hamilton, DeMorgan, Jevons, and even Bradley) devote a great deal of time and space to a discussion of Newton's methodological ideas. More importantly, their philosophical doctrines show signs of having come to grips, at a fundamental level, with the epistemological implications of both Newtonian physics and Newtonian philosophy of science.

How is the historian to explain this profound transformation? When, and by whom, was Newton introduced into the mainstream of British philosophical thought on epistemology and the philosophy of science? What happened within British empiricism that caused it to take Newton the methodologist as seriously as it had previously taken Newton the scientist? I submit that a partial and tentative answer to all these question will come from a close analysis of the methodological writings of Thomas Reid (1710–96), leader of the Scottish school of common-sense philosophy. Indeed, most of the available evidence seems to indicate that Reid was the first major British philosopher to take Newton's opinions on induction, causality, and hypotheses seriously.

Of all the British philosophers of the 18th century, Reid was ideally suited to serve as Newton's spokesman. Trained in the natural sciences, Reid lectured at Aberdeen and Glasgow in physics and mathematics as well as philosophy in the narrower sense. His first publication was a defence of Newtonian mechanics in the *Philosophical Transactions.*[6] A close friend of the Newtonians James Stewart and David Gregory, and the nephew of James Gregory, he was a devout partisan of Newtonian physics and gave lectures on *Principia* for some twelve years. (It is significant that the first section of Reid's surviving lecture notes on natural philosophy – written by one of his students – was devoted to Newton's *Regulae Philosophandi*; maxims which, as we shall see, were later to play so important a role in Reid's philosophical work.) Reid, in short, was a well-read, capable natural philosopher who knew Newton's work first-hand and could see that there was more in it of logical and epistemological interest than the popularized works of Newtoniana might suggest.[7] But if Newton was Reid's scientific inspiration, it was Hume who seems to have stirred his first serious philosophical interests.[8] Reid evidently read the *Treatise* and the *Enquiry* avidly for he wrote to Hume in 1763: "I shall always avow myself your disciple in metaphysics. I have learned more from your writings in this kind, than from all others put together."[9]

However, his self-avowed debt to Hume was a curious one; for almost all of Reid's philosophical works are concerned to show that the Humean scepticism with regard to the senses and the predilections of common sense (as well as Hume's critique of the sciences) is without foundation. Reid believed that the Humean account of knowledge, causality, and induction was so far-fetched as to constitute a *reductio ad absurdum* of the basic premises of classical empiricism. He felt it was necessary to start again at the foundations in order to give an account of knowledge and the means of obtaining it which distinguished various degrees of certainty and reliability in our knowledge. As Reid interpreted Hume, empiricism was no longer able to distinguish between the merits of, say, the astrologers and the classical mechanicians and therefore was no longer appealing as a theory of knowledge.[10] While granting Hume's point that no empirical knowledge was infallible, he maintained that there were degrees of fallibility and he insisted that an epistemology and philosophy of science should offer guidelines for deciding between reliable and unreliable statements or systems. In constructing such an epistemology, Reid drew extensively on the insights which he had gleaned from his lengthy familiarity with the writings of Newton. It is important to point out, however, that Reid had more than one reason for taking Newton seriously; for he saw his own work as an attempt to create a scientific mental philosophy, that is, psychology.[11] He continually preached that the mental sciences (viz., psychology and epistemology) were so backward precisely because no one had tried to construct them according to the canons of scientific evidence and proof. What finer aim could one have, he reasoned, than to seek to structure the philosophy of mind along scientific lines, and this, of course, meant trying to shape the mental sciences so as to conform to Newton's methodological insights.

THE POLEMIC AGAINST HYPOTHESES

One of the foundation stones of Reid's philosophical system and the central attitude he adopted from Newton, was his suspicion of, bordering on contempt for, any theories, hypotheses, or conjectures which are not *induced* from experiments and observations. He maintained that a patient and methodical induction coupled with a scrupulous repudiation of all things hypothetical was the panacea for most of the ills besetting philosophy and science. Thus, in his *Essay on the Intellectual Powers of Man*, he writes:

> [Scientific] discoveries have always been made by patient observation, by accurate experiments, or by conclusions drawn by strict reasoning from observations and experiments, and such discoveries have always tended to refute, but not to confirm, the theories and hypotheses which ingenious men have invented. As this is a fact confirmed by the history of philosophy in all past ages, it ought to have taught man, long ago, to treat with contempt *hypotheses in every branch of philosophy, and to despair of ever advancing real knowledge in that way.*[12]

Elsewhere, he notes that "philosophy has been, in all ages, adulterated by hypotheses; that is, by systems built partly on facts, and much upon conjecture" (*Intellectual Powers*, p. 249). Reid was particularly antagonistic to hypotheses when discussing his program for transforming the philosophy of mind into a science. "Let us", he insists,

> lay down this as a fundamental principle in our inquiries into the structure of the mind and its operations – that no regard is due to the conjectures or hypotheses of philosophers, however, ancient, however generally received (*Intellectual Powers*, p. 236).

Reid has several objections to the hypothetical method, some logical, others historical or psychological:

(1) *As a matter of historical fact, hypotheses and conjectures have not been very productive, and have tended to mislead rather than enlighten us.* Despite Reid's familiarity with the history of science (he would probably have said because of it), he maintained that there was not a single law or discovery which was the result of conjecturing about nature. He argued that if even *one* "useful discovery" could be credited to the hypothetical method then

> Lord Bacon and Sir Isaac Newton have done great disservice to philosophy by what they have said against hypotheses. But, if no such instance can be produced, we must conclude, with those great men, that every system which pretends to account for the phenomena of Nature by hypotheses or conjecture, is spurious and illegitimate, and serves only to flatter the pride of man with a vain conceit of knowledge which he has not attained (*Intellectual Powers*, p. 250).

(2) *The adoption of an hypothesis prejudices the impartiality of the scientist.* In so far as a conjecture or hypothesis is a creation of our own minds, we tend to take a vested interest in it and are less eager to test it severely.[13] Reid argues that science floundered so long in the Middle Ages precisely because the Galenic hypotheses in medicine, the Ptolemaic hypotheses in astronomy, and the Aristotelian hypotheses in physics dissuaded the Schoolmen from making the sorts of experiments which eventually led to the downfall of these systems.

Moreover, once we discover an hypothesis that is appealing on *a priori* grounds, we are prone to decide the truth or falsity of empirical propositions by their accord with that hypothesis rather than by testing our hypothesis against empirical statements whose truth has been determined independently.[14] Almost a century earlier, Newton himself had been forced to develop an argument very like Reid's in his early optical controversy with Hooke,

Huygens, and Pardies.[15] Reid points out that preconceived hypotheses even tend to influence the way in which we interpret observations:

A false system onced fixed in the mind, becomes, as it were, the medium through which we see objects: they receive a tincture from it, and appear of another colour than when seen by a pure light (*Intellectual Powers*, p. 474).

(3) *The hypothetical method presupposes a greater simplicity in nature than we find there*. In any given domain of science or philosophy, there are certain obvious phenomena which require explanation. If we indulge in hypotheses, and if we are sufficiently ingenious, we can always find several different hypotheses which will "save" or explain those phenomena. The most common way of choosing between such explanatory hypotheses is to select the simplest of the set. But, and here Reid uses the historical argument again, the development of the sciences shows conclusively that those hypotheses that seem the simplest are generally quite wide of the mark. Nature, in his view, is highly complex and "correct" theories like Newton's only seem simple in retrospect because they are sufficiently complex to match nature's intricacy. He particularly objects to the tendency of many philosophers to give *a priori* preference to a system just because its structure is simple or its principles few.[16] Even Newton comes in for criticism here since he was, on Reid's view, sometimes "misled by analogy and the love of simplicity" (*Human Mind*, p. 207; cf. *Intellectual Powers*, p. 471).

(4) *The use of hypotheses assumes that man's reason is capable of understanding the works of God*. In Reid's view, nature is complex because it is the handiwork of a Being whose wisdom and ingenuity are infinitely greater than man's. If we were presented with a machine or other contrivance (say a clock) made by another human being, then it would not be unreasonable, prior to our empirical inspection of its clockwork, to conjecture about its internal arrangement. Because its maker is no more ingenious than we are, we might have a reasonable chance of guessing the correct arrangement. But,

The works of men and the works of God are not of the same order. The force of genius may enable a man perfectly to comprehend the former, and see them to the bottom. What is contrived and executed by one man may be perfectly understood by another man ... But the works of nature are contrived and executed by a wisdom and power infinitely superior to that of man; and when men attempt, by the force of genius, to discover the causes of the phenomena of Nature, they have only the chance of going wrong more ingeniously. Their conjectures may appear very probable to beings no wiser than themselves; but they have no chance to hit the truth.[17]

(5) *Hypotheses can never be proved by 'reductio' methods*. Although

proponents of the hypothetical method were aware of the fallacy of affirming the consequent, and generally conceded that hypotheses could not be proved by comparing their consequences with observation, many of them maintained that an hypothesis could be proven indirectly by a series of crucial experiments.[18] Newton was among the first to point out the fallacy in this reasoning,[19] by showing the unlikelihood of being able to enumerate all the possible hypotheses for explaining a class of events and Reid follows him in dismissing the notion that hypotheses can be proved in this way: "This, indeed, is the common refuge of all hypotheses, that we know no other way in which the phenomena may be produced, and therefore, they must be produced in this way" (*Intellectual Powers*, p. 250).

(6) *The use of hypotheses usually violates Newton's first "Regula Philosophandi"*. Newton's first rule – that no more causes are to be admitted than those which are both true and sufficient to explain the appearances – was often used by Reid, but rarely so effectively as in his critique of the method of hypothesis.[20] That rule, he argued, requires not only that hypotheses be adequate to explain all the appearances, but also that the mechanisms and entities postulated in those hypotheses be true, and not merely fictions of the human imagination. Reid asserts, with some justice, that the "votaries of hypotheses" were generally satisfied if their hypotheses saved the appearances and that they were not concerned to know whether their hypothetical entities and mechanisms corresponded to physical reality. In an important letter to Lord Kames, Reid elaborates on this point and on the meaning of that cryptic Newtonian phrase "true cause":

> All investigation of what we call the causes of natural phenomena may be reduced to this syllogism – If such a cause exists, it will produce such a phenomenon; but that cause does exist: Therefore, &c. The first proposition is merely hypothetical. And a man in his closet, without consulting nature, may make a thousand such propositions, and connect them into a system; but this is only a system of hypotheses, conjectures, or theories; and there cannot be one conclusion drawn from it, until he consults nature, and discovers whether the causes he has conjectured do really exist.[21]

Five years later, and indicating still more clearly his debt to the first rule, Reid wrote:

> When men pretend to account for any of the operations of nature, the causes assigned by them ought, as Sir Isaac Newton has taught us, to have two conditions, otherwise they are good for nothing. *First*, they ought to be true, to have a real existence, and not to be barely conjectured to exist, without proof. *Secondly*, they ought to be sufficient to produce the effect.[22]

Because Reid's interpretation of Newton's doctrine of *vera causa* became highly influential on such writers as Dugald Stewart, John Herschel, and Charles Lyell, it is worth discussing in some detail. As Reid construes the first *regula*, it amounts to the claim that any putative causal explanation (a) must be sufficient to explain the relevant appearances and (b) must postulate entities and mechanisms whose existence can be *directly* ascertained. Condition (b) is the crucial one because it is meant to explicate the demand for 'true causes'. What this amounts to is the claim that *unobservable entities*, because we can have no *direct* evidence of their existence, *have no role to play in causal explanations*. In Reid's hands, Newton's first rule of reasoning becomes a vehicle for excluding all theoretical entities from natural philosophy. The only causes which it is legitimate to postulate are ones which, like the events they are invoked to explain, can be observed and measured. Like Hume before him (see the preceeding chapter), Reid will not traffic in theories whose predicates are not drawn directly from sense.

Put differently, Reid uses Newton's *vera causa* doctrine as a tool for discrediting the argument – usually associated with the method of hypothesis – that unobservable causes can be legitimately invoked so long as their existence can be *indirectly* confirmed,[23] i.e., so long as they are capable of explaining what can be observed. Reid's position is that no amount of confirmation or explanatory success is sufficient to establish the existence of an unobservable entity. Such tests at best establish the bare *possibility* that such entities exist, but no tests of this kind can conceivably warrant the claim that such entities really exist. Reid puts it this way:

> Supposing it [viz., a certain hypothesis] to be true, it affirms only what may be. We are, indeed, in most cases very imperfect judges of what may be. But this we know, that were we ever so certain that a thing may be, this is no reason for believing that it really is. A *may-be* is a mere hypothesis, which may furnish matter for investigation, but is not entitled to the least degree of belief. The transition from what may be to what really is, is familiar and easy to those who have a predilection for a hypothesis . . . .[24]

(7) *The method of hypothesis substitutes premature theoretical ingenuity for painstaking experimental rigour.* As a general rule, inductivists have been suspicious of the sharp mind and the quick wit of genius. They have sought to reduce scientific discovery to a mechanical process, where differences of intelligence have little or no role to play and where, to use Bacon's phrase "all wits and understandings [are] nearly on a level".[25] Bacon, Newton, and Mill – the classic English inductivists – all shared the belief that a 'logic of discovery' should establish rules for discovering scientific laws and theories

and that these rules should be so nearly mechanical and foolproof that there would be little need for a facile mind or a fertile imagination.[26] Reid unreservedly sides with this point of view. "The experience of all ages", he maintains,

shows how prone ingenious men have been to invent hypotheses to explain the phenomena of nature; how fond, by a kind of anticipation, to discover her secrets . . . . Instead of a slow and gradual ascent . . . by a just and copious induction, they would shorten the work, and by a flight of genius, get to the top at once.[27]

Genius and the creative intellect have only a very circumscribed role to play in science, properly understood:

[genius] may combine, but it must not fabricate. It may collect evidence, but it must not supply the want of it by conjecture. It may display its powers by putting Nature to the test in well-contrived experiments, but it must add nothing to her answers (*Intellectual Powers*, p. 472).

One of the severe limitations of the method of hypothesis is that it relies too heavily on the ingenuity of the theoretician to second-guess nature, to anticipate the outcome of unperformed experiments. Given Reid's opinions on the complexity of nature and the limitations of man's imagination compared with God's, it is clear why he held no hope that genius unaided by extensive experimentation could advance very far. So adamant was Reid on this point, that he even chided Newton for his occasional lapses into the language of hypothesis. After pointing out several errors in the hypothesis in the fifteenth query to the *Opticks* (regarding the transmission of images from the eye to the brain), Reid concludes that even "if we trust to the conjectures of men of the greatest genius in the operations of nature, we have only the chance of going wrong in an ingenious manner".[28] Feeble as an instrument of discovery, and therefore impotent as a probe for truth, the method of hypothesis emerges from Reid's attack with very little to recommend it.

It was not only hypotheses that Reid found objectionable; theories, too, were anathema to him. Indeed, he tended to conflate the terms 'theory', 'hypothesis', and 'conjecture' by attacking them indiscriminately.[29] He leaves us in no doubt that it was from Newton that he inherited his aversion to all things hypothetical: "I have, ever since I was acquainted with Bacon and Newton, thought that this doctrine [i.e., never to trust hypotheses and conjectures] is the very key to natural philosophy, and the touchstone by which everything that is legitimate and solid in that science, is to be distinguished from what is spurious and hollow . . . ."[30] Elsewhere, he speaks of the example "the great Newton" set in the *Opticks*, an example which "always ought to be but rarely hath been followed" (*Human Mind*, p. 180), whereby he distinguishes his speculations (which he put "modestly" in the Queries) from his indubitable conclusions.[31] It was, Reid tells us, precisely because Newton took such care to distinguish his hypotheses from his inductively-established theories that he "considered it as a reproach when his system was called his hypothesis; and says, with disdain of such imputation, *Hypotheses non fingo*" (*Intellectual Powers*, p. 250).

Like Newton, Reid maintained that there was a crucial difference in kind between the laws and properties discovered by induction and the hypotheses and theories devised to explain those laws and properties. Moreover, the same kind of vacillation that scholars have detected in Newton's writings[32] between an explicit denunciation of all hypotheses and a grudging admission of the possible utility of some hypotheses is echoed by a similar ambiguity in Reid's position. At one moment, he tells us that hypotheses should be "treated with contempt", that they are "apocryphal and of no authority" and that they "cannot produce knowledge"; but when pressed on this point he concedes that "conjecturing may be a useful step even in natural philosophy"[33] and that hypotheses "may lead to productive experiments".[34] Nonetheless, the thrust of almost all of Reid's arguments is towards discrediting those philosophers (e.g., David Hartley) and scientists (e.g., the Cartesians) who did use hypotheses, however sparingly.

Fundamentally, what seems to have disturbed Reid more than anything else about the method of hypothesis was the subtle way in which, so far as he was concerned, it distorted and perverted the true aims of science. For Reid, the chief goal was the discovery of how nature operated and a faithful description of those operations. But once the natural philosopher was allowed to indulge his fancy with hypotheses, such hypothesizing became the chief aim, to which all else was subordinate. The situation had become so bad that Reid was forced to observe that "the invention of a hypothesis, founded on some slight probabilities, which accounts for many appearances of nature, has been considered as the highest attainment of a philosopher".[35] Reid, of course, was not alone in interpreting Newton's "hypotheses non fingo" literally and in applying it even more broadly than Newton had done. It was a characteristic of the time to lampoon hypotheses and their proponents and to treat them as anachronistic legacies of the pre-Newtonian era. As Benjamin Martin noted shortly after the appearance of Reid's *Human Mind*:

The Philosophers of the present Age hold [hypotheses] in vile Esteem, and will hardly admit the Name in their Writings; they think that which depends upon bare Hypothesis and Conjecture, unworthy the name of Philosophy; and therefore have framed new and more effectual Methods for philosophical Enquiries.[36]

Reid was, however, the first of his philosophical colleagues to seek to build an entire epistemology around Newton's methodological *aperçus*.

In a sense, Reid's conception of an 'hypothesis' and his rather shabby treatment of the hypothetico-deductive method are logical extensions of the trend which was started by Newton and which reached its culmination in the work of Mill. By gradual stages, the meaning of the term hypothesis was eroded and altered, much to the confusion of subsequent methodological inquiry. In the early 17th century, 'hypothesis' had signified any general proposition which was assumed, but not known, to be true. It was used especially to refer to the unproven postulates, axioms, or first principles of any science, and this meaning of the term had been common since Aristotle and Euclid. Even Newton used 'hypothesis' in this sense in the first edition of *Principia* (1687) and it carried no particularly pejorative connotations there.[37] But the signification of the term was gradually transformed in Newton's later writings. In a constant battle with the Cartesians, Newton would often find his opponents offering theories or conjectures which were patently false when tested empirically. Thus, six of Descartes' seven laws of motion were obviously incompatible with the most elementary impact experiments. Similarly, Descartes' vortex hypothesis was demonstrably false, even if its refutation was rather more elaborate than in the case of the laws of motion. Undeterred by such anomalies, many Cartesians continued to hold to variations of Descartes' laws of motion and vortex theory, defending them in terms of their *a priori* cogency. Understandably, Newton had no patience with such an approach and he tried to discredit it with methodological arguments. The "a priorists", he insisted, were using hypotheses; so the natural way to eliminate their non-empirical techniques was to insist, as he finally did, that "hypotheses have no place in experimental philosophy".[38] Unfortunately, Newton's blanket denunciation of hypotheses not only discredited the "a priorists", it also left no place for an *empirical* hypothetico-deductive method, which insisted on the experimental testing of all conjectures. What tended to happen, therefore, was that the older meaning of 'hypothesis' (as an axiom or postulate) came to be confounded with the notion of an unempirical or untestable proposition. Where the two had been clearly distinguished, they were now indiscriminately confused and the legitimate arguments against *a priori* and untestable hypotheses were mistakenly used against all hypotheses whatever.[39] By Reid's time, as we have seen, the method of hypothesis was in such ill repute that Reid, although perfectly able to distinguish the two senses of 'hypothesis', considered them equally objectionable and demanded that both were to be scrupulously avoided.[40] Indeed, so vague had the term 'hypothesis' become that Reid could include 'theory' in its connotation and could claim that Newton's arguments against hypotheses were equally arguments against theorizing *tout court*!

In assessing the propriety, let alone the validity, of Reid's attack on hypotheses, one must take certain mitigating circumstances into account. Reid was concerned, above all else, with founding a science of mind or 'pneumatology', as he called it. He took methodological questions so seriously because he felt it was necessary to see that the new science had adequate criteria for explanation and sound guide-lines for investigation. Comparing the fledging mental science with natural science, one thing was evidently clear to him: although hypotheses in physics could easily be detected and exposed, hypotheses dealing with mental phenomena were not so easily disposed of. To take an elementary example, if one adopts the conjecture that the mind at birth is a *tabula rasa*, there is no clear-cut way of confirming or refuting such an hypothesis. Reid, sensitive to the extent to which hypotheses about mental events were insulated from observable phenomena, took the extreme course of insisting that hypotheses must be consistently rejected.

One must remember, too, that Reid could well afford to be so cavalier about hypotheses and so indiscriminate in his depreciation of them. Reid, after all, thought he had a method which dispensed with hypotheses once and for all – the method of induction.

INDUCTION AND THE UNIFORMITY OF NATURE

Before we examine the origins and structure of Reid's views on induction, a word should be said about "Newtonian" and "Baconian" induction; otherwise, it will be difficult to decide whether Reid's theory of induction derives from one or the other. At the outset, it must be noted that Reid writes as if the Baconian and Newtonian theories of induction amount to the same thing and he is continually paying lip-service to both. Generally, he maintains that Bacon developed the theory of induction and that Newton subsequently applied it in a brilliant fashion to astronomy and optics.[41] Newton, moreover, simplified and codified the principles of Baconian induction in his *Regulae Philosophandi*. This might lead one to believe that Reid was a Baconian essentially and only incidentally a Newtonian methodologist. But to take such a view would be a serious mistake. Baconian induction was concerned with the discovery of *forms* by a study and classification of *simple natures*. Bacon's method was essentially the Aristotelian one of definition by genus-species relationships. It was not so much a method of determining laws as it was a technique for generating definitions. However, with Newton, as with Reid, induction is conceived in an entirely different mode. In their work, there is no analogue for Bacon's forms or simple natures; nor is their induction meant to be foolproof, as Bacon's certainly was. Moreover, the method of elimination, which was such an integral part of Bacon's scheme, is repudiated by both Newton and Reid.[42] Accordingly, all the evidence I can find points to this conclusion: Reid was only vaguely familiar with the details of Bacon's *Novum Organum* and he assumed, as was common at that time, that Newton (whose work he did know well) got his ideas on induction from Bacon and that Newton's writings on method were enlargements on, but perfectly compatible with, Bacon's views.[43]

Basically, Reid's inductive method had three components: (1) observation of facts and experimentation; (2) "reduction" of these facts to a general rule or law; and (3) the derivation of further conclusions from the set of general rules.[44] These three steps embody, so far as Reid is concerned, the essence of Newton's basic doctrines,[45] and Reid candidly concedes that this is "the view of natural philosophy which I have learned from Newton."[46] The crucial point, of course, is the second step and Reid is as vague on this score as most of his fellow inductivists have been.[47] He does not, by any means, believe that we can legitimately generalize every set of similar observations. Like Bacon, he insists that the amount of data should be large and that our generalizations must be gradual. Indeed, he argues that the only real difference between genuine science and pseudo-science is that in the former, our observations are more numerous and our generalizations less hasty:

> Omens, portents, good and bad luck, palmistry, astrology, all the numerous arts of divination and interpreting dreams, false hypotheses and systems, and true principles in the philosophy of nature are all built upon the same foundation in the human constitution, and are distinguished only according as we conclude rashly from too few instances, or cautiously from a sufficient induction (*Human Mind*, p. 113).

So far as one can tell from Reid's text, he usually uses 'induction' in the straightforward enumerative sense of generalization from particulars to universals. Whatever properties are found to belong to all the examined individuals of a particular type are presumed to belong to all instances of that type. This kind of induction, from particular facts to general conclusions, is "the master-key to the knowledge of Nature, without which we could form no general conclusions . . . " (*Intellectual Powers*, p. 402). All we can know in natural science is what we can learn by induction:

In the solution of natural phenomena, all the length that the human faculties can carry us, is only this, that from particular phenomena, we may by induction, trace out general phenomena, of which the particular ones are necessary consequences.[48]

But Reid never suggests any criteria for distinguishing between what he considers a *just* or *sufficient* induction and a premature generalization. On the other hand, he would probably have defended himself against the charge that he was needlessly vague on the meaning of induction by pointing out that Bacon and Newton had already described the nature and conditions of inductive inference and that he himself simply acquiesced in their analysis. Thus, of Bacon, he writes: "The rules of inductive reasoning, or of a just interpretation of Nature, have been, with wonderful sagacity, delineated by the great genius of Lord Bacon: so that his *Novum Organum* may be justly called 'A Grammar of the Languge of Nature' " (*Human Mind*, p. 200). Again he observes: "Lord Bacon first delineated the strict and severe method of induction; since his time it has been applied with very happy success in some parts of natural philosophy and hardly in anything else."[49]

Notwithstanding his high regard for Bacon, Reid insisted that it was Newton's rules of philosophizing which codified and embodied Bacon's inductive principles and it was to them that Reid invariably turned when dealing with questions of scientific method.[50] Newton's *regulae*, he maintains, "are maxims of common sense, and are practiced everyday in common life; and he who philosophizes by other rules, either concerning the material system or concerning the mind, mistakes his aim" (*Human Mind*, p. 97). Indeed, all of Reid's writings, both published and unpublished, are teeming with laudatory references to Newton's rules of philosophizing; and in this case, Reid's deferential lip-service, which is misleading when paid to Bacon, is indicative of a profound debt on Reid's part to those four propositions at the beginning of the third book of the *Principia*. He called them "the only solid foundation on which natural philosophy can be built" (*Intellectual Powers*, p. 436), and argued that Newton's physics "would never be refuted because it was based on these self-evident principles" (ibid.). There is a particularly interesting fascicule among Reid's MSS. which illustrates his high regard for the *regulae*. It is titled "Of the Order in Which Natural Philosophy Ought To Be Taught". Of the nine headings in his outline, the second is: "Laws of Philosophizing from S^{r}. Is. Newton's Princ. Lib. 3".[51] Again, in his lecture notes on pneumatology (*c*. 1768–69), he calls the *regulae* "the axioms upon which men reason in physicks".[52] In an unpublished draft version of the *Essays on the Intellectual Powers of Man*, he writes:

there are also first principles in physicks or natural philosophy. Sir Isaac Newton has laid down some of the most important of these in the third book of his *Principia* under the name of axioms or laws of philosophizing, & by this means has given a stability to that science which it had not before.[53]

Just as the rules of the syllogism govern deductive reasoning, so must Newton's *regulae* be the criteria for assessing the validity of all empirical or inductive reasoning.[54] Although Reid concedes that the *regulae* are not, as the laws of syllogistic are, *a priori*, he argues that they are nonetheless so obvious to men of good sense that they can be used legitimately as canons for all scientific and philosophical research.[55] Besides copious allusions to Newton's *regulae*, Reid also adopts the analysis-synthesis language of Newton's optical queries when discussing questions of scientific method. In an almost literal transcription of a section from Query XXXI, Reid writes to Kames:

Our senses testify particular facts only: from these we collect by induction, general facts, which we call laws of nature of natural causes. Thus, ascending by a just and copious induction, from what is less to what is more general, we discover as far as we are able, natural causes or laws of nature. This is the analytic part of natural philosophy. The synthetical part takes for granted, as principles, the causes discovered by induction, and from these explains or accounts for the phenomena which result from them. This analysis and synthesis make up the whole theory of natural philosophy . . . From this view of natural philosophy, which I have learned from Newton, your Lordship will perceive that no man who understands it will pretend to demonstrate any of its principles . . .[56]

Despite his high regard for induction as a method, Reid – again like Newton – was quite prepared to admit its fallibility.[57] No matter how cautious our inductions and regardless of how extensive our evidence, we may discover that some of the propositions we think to be natural laws are, in fact, false.[58] However keenly Reid might have wanted to establish Newton's *regulae* as indubitable first principles, no post-Humean philosopher could altogether neglect the doubts that the *Enquiry* and the *Treatise* had cast on the legitimacy of induction and the assumption of the uniformity of nature. Indeed, Hume's analysis seemed to undermine both the second[59] and the third[60] rules of philosophizing. Reid saw it as one of his tasks to defend those *regulae* (which were, after all, "the foundations of the philosophy of common sense") from Hume's onslaughts.

Hume had argued that our belief in the continuance of nature's laws could not be derived from reason, that we could neither know nor even rationally expect that the future will resemble the past. In his view, our confidence in the uniformity of nature was merely a belief or expectation, to be distinguished sharply from perception, memory, and imagination. Hume distinguished those four in terms of the vivacity or faintness of the associated idea. Suppose, for instance, that the idea is one of a sunrise. A perception of a sunrise here and now is very vivid; a memory of a sunrise is fainter; a belief that the sun will rise tomorrow is fainter still; and our imagining a sunrise is so faint and lacklustre as to carry no conviction of its truth at all. This attempt to distinguish perception, memory, prophecy or prediction, and imagination solely in terms of the vividness of the accompanying ideas, is rejected by Reid. He points out that if one takes a memory of yesterday's sunrise, and slightly fades its vivacity, one obtains, not an idea of future sunrise, but an idea of a sunrise in the more distant, but remembered, past. Surely, he argues, it is absurd to suggest that "there is a certain period of this declining vivacity, when, as if it had met an elastic obstacle in its motion backward, it suddenly rebounds from the past to the future, without taking the present in its way" (*Human Mind*, p. 198).

Notwithstanding Reid's rejection of Hume's distinction between ideas of past, present, and future in terms of their vivacity, he accepts Hume's thesis that our belief in the uniformity of nature is instinctive and non-rational. But on Reid's view, it is instinctive not by virtue of the habitual associations of experience but because of an innate disposition in man's nature, which Reid calls *the inductive principle*:

> ... our belief in the continuance of nature's laws is not derived from reason. It is an instinctive prescience of the operations of nature, very like to that prescience of human actions which makes us rely upon the testimony of our fellow creatures ... Upon this principle of our constitution, not only acquired perception but also inductive reasoning, and all our reasoning from analogy, is grounded; and therefore, for want of another name, we shall beg leave to call it *the inductive principle*. (*Human Mind*, p. 199)

The disposition to believe in the uniformity of nature is innate and not, as Hume suggested, a habit acquired from experience. "Prior to all reasoning", Reid insists, "we have, *by our constitution*, an anticipation that there is a fixed and steady course of nature ... " (ibid.; my italics). A young child does not have to be burned more than once before he keeps a healthy distance from the fire. Indeed, experience, far from giving rise to the inductive principle, as Hume suggests, tends to circumscribe it and makes us more critical and restrained in following its dictates: "This [inductive] principle, like that of credulity, is unlimited in infancy, and [becomes] gradually restrained and regulated as we grow up" (ibid.). The important point, however, is that man's mind is structured in such a way that he inevitably presupposes uniformity in nature:

We are so made that, when two things are found to be conjoined in certain circumstances, we are prone to believe that they are connected in nature, and will always be found together in like circumstances. The belief which we are led into in such cases is not the effect of reasoning, nor does it arise from intuitive evidence in the thing believed; it is, as I apprehend, the immediate effect of our constitution" (*Intellectual Powers*, p. 332; cf. p. 451).

To ask, as Hume did, about the justification for our belief in the uniformity of nature, is to misconceive the nature of the inductive principle. It is simply instinctive, constitutional, and innate; like Dr Johnson, Reid felt that we can no more justify it than we can rid ourselves of it. Reid sees the inductive principle, construed as a disposition to belief, as the ground for, and the justification of, Newton's second *Regula*: "It is from the force of this principle that we immediately assent to that axiom upon which all of our knowledge is built. That effects of the same kind must have the same cause . . . ."[61]

It is true that we can have no conception of the uniformity of nature prior to our having experiences or sensations, and in this sense experience is a necessary condition for our belief in the uniformity of nature. But – and here Reid's reaction to Hume resembles Kant's – our belief in the uniformity of nature and in the causal axiom *is not derived from experience*. Without the inductive principle, we could learn nothing from nature because generalization would be impossible[62] and science, either natural or mental, would be unthinkable.[63]

Reid not only maintains that a belief in the uniformity of nature is perfectly natural and inevitable, he also insists that nature itself is uniform and that every change in the material system is strictly governed by immutable laws. He claims that God is the author of all things and that, in his wisdom, he made a univers governed by fixed laws. We have no way of knowing with certainty whether the propositions we think to be laws are, in fact, the true laws of the universe, and that is why our inductions must be cautious and our conclusions always subject to modification. However, man's fallibility is certainly no sign that the universe is governed by chance and Reid has no qualms about stating categorically that all physical events are subject to scientific laws. None of Reid's arguments really constitutes a satisfactory answer to Hume's doubts, and one is almost inclined to say that he simply took the positions which Hume found most dubious (e.g., "nature is uniform") and asserted them as first principles or as matters of faith.

CONCLUSION

In discussing some of the salient features of Reid's philosophy of science, I hope I have been able to convey some idea of the pervasive and fundamental Newtonian influences to be found in his work. Unlike the older empiricists, who had neither the scientific background nor the philosophical biases necessary for a sympathetic reading of Newton, Reid was able to draw equally on the insights of the empiricists and the inductivists and to fuse the two, thereby giving rise to the empiricist-inductivist tradition that was to play so large a role in 19th-century philosophy of science.

NOTES

1 David Hume, *A Treatise of Human Nature*, ed. L. A. Selby-Bigge (Oxford, 1896), p. xx. Newton himself had suggested that his new methods might well have a bearing on moral philosophy: "And if natural philosophy in all its parts, by pursuing this Method shall at length be perfected, the Bounds of Moral Philosophy will also be enlarged." See Sir Isaac Newton, *Opticks* (New York, 1952), Query XXXI. Taking up this theme, Pope insisted that we should "Account for Moral, as for nat'ral things." See Alexander Pope, *An Essay on Man* (London, 1786), Epistle IV.

2 In each of these domains, and others as well, one can point to several figures who wanted to "Newtonize" their subject in various ways. The most overt of these are the attempts to model works of politics or theology strictly along the deductive lines of the *Principia*. For example, George Cheyne in part 2 of his *Philosophical Principles of Religion: Natural and Revealed* (London, 1715) constructs an elaborate theological system beginning with definitions and axioms and then proceeding to a series of theorems, lemmas, and corollaries deduced from the axioms.

3 Even a brief sampling of 18th-century texts illustrates how enthusiastically physicists adopted Newton's call for a non-conjectural science. Thus, Oliver Goldsmith described how science had progressed from being an "hypothetical system" to an "authentic experimental system" (*A Survey of Experimental Philosophy* [London, 1776], p. 4). W. 'sGravesande, an eminent Dutch Newtonian, observed that, thanks to Sir Isaac, "all hypotheses are to be rejected" in natural philosophy (*Mathematical Elements of Physics, Proved by Experiments* [London, 1720], I, 5). 'sGravesande went on to insist that "He only, who in physics reasons from phenomena, rejecting all feigned hypotheses, and pursues this method to the best of his power, endeavors to follow the steps of Sir Isaac Newton, and very justly declares that he is a Newtonian philosopher . . . " (ibid., p. xi). Henry Pemberton, in his *View of Sir Isaac Newton's Philosophy* (London, 1728), contrasts the unreliable hypothetical method which "makes a hasty transition from our first and slight observations on things to general axioms" with the Newtonian and

Baconian method of cautious ascent (p. 5). Writing about the same time, 1731, the Dutch physicist Musschenbroek asserted that as a result of Newton's work, "all hypotheses are banned from physics" (cited in Rosenberger's *Geschichte der Physik* [Braunschweig, 1887], III, 3). Earlier still, one of Newton's first followers, George Cheyne, asserted that " . . . imaginary or *Hypothetical* Causes, have no place in true Philosophy" (above, n. 2, pt. 1, p. 45).

Perhaps the most explicit claim as to the infallibility of Newton's physics was Emerson's observation that "It has been ignorantly objected by some that the Newtonian philosophy, like all others before it, will grow old and out of date, and be succeeded by some new system . . . But this observation is falsely made. For never a philosopher before Newton ever took the method that he did. For whilst their systems are nothing but hypotheses, conceits, fictions, conjectures, and romances, invented at pleasure and without any foundation in the nature of things. [*sic*] He, on the contrary . . . admits nothing but what he gains from experiments and accurate observations . . . It is therefore a mere joke to talk of a new philosophy . . . Newtonian philosophy may indeed be improved and further advanced; but it can never be overthrown; not withstanding the efforts of all the Bernoulli's, the Leibnitz's . . . " (*Principles of Mechanics* [London, 1773], n. p.).

4 Many historians of philosophy have overlooked the fact that the major sources of Newton's methodological views were not available when Locke's *Essay* was written. This has led to two serious errors: (1) the suggestion that Locke, as a presumed Newtonian, was a vehement opponent of hypotheses in science, and (2) the claim that Locke's general theory of method, including his supposed aversion to hypotheses, was derived from Newton. For a discussion of these questions, see Chapter 5.

5 Cf. K. Popper's 'A Note on Berkeley as a Precursor of Mach', *British Journal for the Philosophy of Science* 5 (1953), 26–36, and G. J. Whitrow's 'Berkeley's Philosophy of Motion', ibid., 37 ff.

6 For a discussion of, and excerpts from, Reid's paper in the *Phil. Trans.*, see my 'Postmortem on the *Vis Viva* Controversy', *Isis* 59 (1968), 131–43.

7 Few intellectuals, besides trained mathematicians, could read *Principia*, or even the *Opticks*, without great difficulty. As a result popularized works were much in vogue, such as Voltaire's *Elements of Newton's Philosophy*, Pemberton's *View of Sir Isaac Newton's Philosophy* (London, 1728), and Maclaurin's *Account of Sir Isaac Newton's Philosophical Discoveries* (London, 1750). Although these books all praised Newtonian science for the certainty of its conclusions and the experimental bias of its founder, none of them did justice to Newton's theory of scientific method. For a discussion of some of these works, see I. B. Cohen's *Franklin and Newton* (Philadelphia, 1956).

8 In this as in other respects, the intellectual careers of Reid and Kant are much alike.

9 Reid to Hume, 18 March, 1763, *Works of Thomas Reid, D. D.*, ed. Hamilton, 6th ed. (Edinburgh, 1863), I, 91. Hereafter all page references will be to the *Works* unless indicated otherwise.

10 This reading of Hume as a total sceptic, unable to distinguish sound from unsound judgment, is not as far-fetched as it might seem. After all, it was Hume who wrote "The intense view of these manifold contradictions and imperfections in human reason, has so wrought upon me, and heated my brain, that I am ready to reject all belief and reasoning, and can look upon no opinion even as more probable or likely than another."

11 " . . . if ever our philosophy concerning the human mind is carried so far as to deserve

the name of science, which ought never to be despaired of, it must be by observing facts, reducing them to general rules, and drawing just conclusions from them," *An Inquiry into the Human Mind* (1765), abbreviated to *Human Mind*, in *Works*, I, 122.

12 *Essays on the Intellectual Powers of Man* (1785), hereafter abbreviated to *Intellectual Powers*, in *Works*, I, 235.

13 "When a man has, with labour and ingenuity, wrought up an hypothesis into a system, he contracts a fondness for it, which is apt to warp the best judgment" (*Intellectual Powers*, p. 250). "When a man has laid out all his ingenuity in fabricating a system, he views it with the eye of a parent; he strains phenomena to make them tally with it, and makes it look like the work of nature" (*Intellectual Powers*, p. 472).

14 "The facts are phenomena of . . . nature, from which we may justly argue against any hypothesis, however generally received. But to argue from a hypothesis against facts, in contrary to the rules of true philosophy" (*Human Mind*, p. 132).

15 Newton wrote to Oldenburg: "For the best and safest method of philosophizing seems to be, first diligently to investigate the properties of things and establish them by experiment, and then seek hypotheses to explain them. *For hypotheses ought to be fitted merely to explain the properties of things and not attempt to determine them* . . . " (Turnbull, ed., *Correspondence of Isaac Newton* [Cambridge, 1959], I, 99, my italics). See also Newton's fourth *regula philosophandi*.

16 Thus he writes: "Men are often led into error *by the love of simplicity which disposes us to reduce things to few principles, and to conceive a greater simplicity in nature than there really* is . . . We may learn something of the way in which nature operates from fact and observation; but if we conclude that it operates in such a manner, only because to our understanding that appears to be the best and simplest manner, we shall always go wrong" (*Intellectual Powers*, pp. 470–71). Cf. ch. VII of *Human Mind* (p. 206) where Reid similarly notes that: "There is a disposition in human nature to reduce things to as few principles as possible." Cf. this sentiment with Bacon's remark that "the human understanding is of its own nature prone to suppose the existence of more order and regularity than it finds" (*Novum Organum*, Book I, Aphorism xlv).

17 *Intellectual Powers*, p. 472. Earlier, he had written in a similar vein: "Now, though we may, in many cases, form very probable conjectures concerning the works of men, every conjecture we can form with regard to the works of God has as little probability as the conjectures of a child with regard to the works of a man" (ibid., p. 235). "Men," he writes, "only begin to have a true taste in philosophy when they have learned to hold hypotheses in just contempt; and to consider them as the reveries of speculative men, which will never have any similitude to the works of God" (ibid., p. 309).

18 Reid's adversary, David Hartley, had been one of the more important proponents of *reductio* techniques, as well as a vigorous advocate of the hypothetico-deductive method. Cf. especially Hartley's *Observations on Man* (London, 1749), I, ch. I, Prop. v. Hartley's views are discussed in Chapter 8 below.

19 Newton wrote: "I cannot think if effectual for determining the truth to examine the several ways by which the phenomena may be explained, unless there can be a perfect enumeration of all those ways" (Turnbull, above, n. 15, I, 209).

20 Of all Newton's four *regulae*, it was the first that Reid regarded most highly. It is, he argues, "a golden rule; it is the true and proper test by which what is found and solid in philosophy may be distinguished from what is hollow and vain" (*Intellectual Powers*, p. 236). In the light of Reid's opposition to the principle of simplicity, it may seem

curious that he could speak so glowingly of the first *regula* since that rule seems to presuppose the simplicity of nature. On the other hand, as I point out later, Reid has a rather novel interpretation of the meaning of the first rule.

21 Reid to Kames, 16 December, 1780, *Works*, I, 57.

22 *Intellectual Powers*, p. 261. In the same work, he observes that Newton "laid it down as a rule of philosophizing, that no causes of natural things ought to be assigned but such as can be proved to have a real existence. He saw that . . . the true method of philosophizing is this: From real facts, ascertained by observation and experiment, to collect by just induction the laws of Nature, and to apply the laws so discovered, to account for the phenomena of Nature" (Ibid., pp. 271–2).

23 Reid's perennial adversary, David Hartley, had used an argument very like this to establish the existence of the aether. "Let us suppose", he wrote, "the existence of the aether, with these its properties, to be destitute of all direct evidence, still if it [viz., the aether] serves to account for a great variety of phaenomena, it will have an indirect evidence in its favour by this means" (*Observations on Man* [London, 1749], I, 15). Perhaps the most specious argument of this kind was Bryan Robinson's defence of the aether: "This *Aether* being a very general material Cause, without any Objection appearing against it from the Phaenomena, no Doubt can be made of its Existence: For by how much the more general any Cause is, by so much stronger is the Reason for allowing its Existence. The *Aether* is a much more general Cause than our Air: And on that Account, the Evidence from the Phaenomena, is much stronger in Favour of the Existence of *Aether*, than it is in Favour of the Existence of the Air" (*A Dissertation on the Aether of Sir Isaac Newton* [London, 1747], preface, n. p.).

24 *Intellectual Powers*, p. 397. Years earlier, he had written to Kames: "A cause that is conjectured ought to be such that, if it really does exist, it will produce the effect . . . Supposing it to have this quality, the question remains – Whether does it exist or not? . . . If there be no evidence for it, even though there be none against it, it is a conjecture only, and ought have no admittance into chaste natural philosophy" (16 December, 1780, *Works*, I, 57).

25 Francis Bacon, *Works*, ed. Ellis & Spedding (London, 1858), IV, 63.

26 George Turnbull, Reid's teacher, particularly stressed the mechanical character of scientific discovery: "And by the discoveries made in natural philosophy, we know, that no sooner are facts collected, and laid together in proper order than the true theory of the phenomenon in question present itself" (*Principles of Moral Philosophy*, [London, 1740], p. 59).

27 *Intellectual Powers*, p. 472. Earlier he put the point even more candidly: "The world has been so long befooled by hypotheses in all parts of philosophy, that it is of the utmost consequence to every man who would make any progress in real knowledge, to treat them with just contempt, as the reveries of vain and fanciful men, whose pride makes them conceive themselves able to unfold the mysteries of nature by the force of their genius" (ibid., p. 236).

28 *Human Mind*, p. 181. Reid's contemporary, Jean D'Alembert, was similarly sceptical about finding truth by the hit-or-miss method of conjectures and refutations: "It may be safely affirmed that a mere theoretician (*un physicien de cabinet*) who, by means of reasonings and calculations, should attempt to divine the phenomena of nature and who should afterwards compare his anticipations with facts would be astonished to find how

wide of the truth almost all of them had been" (*Mèlanges de Litterature, d'Historie, et de Philosophie* [Amsterdam, 1767], v, 6).

29 Thus, he writes: "Conjectures in physical patters have commonly got the name of *hypotheses*, or *theories*" (*Intellectual Powers*, p. 235). Sir William Hamilton in his *Lectures on Metaphysics* took Reid to task for confusing theories and hypotheses; see especially ibid., p. 120.

30 Reid to Kames, 17 December, 1780, *Works*, I, 56.

31 He makes a similar point when he notes approvingly that: "Sir Isaac Newton, in all his philosophical writings, took great care to distinguish his doctrines, which he pretended to prove by just induction, from his conjectures, which were to stand or fall according as future experiments and observations should establish or refute them" (*Intellectual Powers*, p. 249). In his own work, Reid follows Newton's example, separating "facts" about the human mind from conjectures. He has, for instance, a separate chapter in the *Inquiry* called "Some Queries Concerning Visible Figure Answered", and a chapter in his *Intellectual Powers* on an "Hypothesis concerning the Nerves and Brain". One of Reid's main criticisms of David Hartley's psychological writings was that they mingled hypotheses with facts.

32 For a discussion of this point, see I. B. Cohen's *Franklin and Newton* (Philadelphia, 1956) as well as his more recent publications.

33 Reid to Kames, 16 December, 1780, *Works*, p. 56.

34 Ibid. Elsewhere he makes a similar point: "Let hypotheses be put to any of these uses as far as they can serve. Let them suggest experiments or direct our inquiries; but let just induction alone govern our belief" (*Intellectual Powers*, p. 251).

35 *Intellectual Powers*, p. 235. More than a century earlier, Newton had deplored the preoccupation with hypotheses among his scientific colleagues. He observed that finding a general hypothesis accounting for all the appearances had become "the philosophers' universall topick" (Turnbull, above, n. 15, I, 96–97; cf. I. B. Cohen, ed., *Isaac Newton's Letters and Papers on Natural Philosophy* [Cambridge, 1958], p. 179).

36 B. Martin, *Philosophical Grammar*, 7th ed. (London, 1769), p. 19. Martin was rather more sympathetic to hypotheses than most of his British contemporaries.

37 For a discussion of the evolution of the meaning of 'hypothesis' in Newton's work, see Cohen's *Franklin and Newton.*

38 Newton, *Principia*, General Scholium. Compare this with Reid's remark that "hypotheses ought to have no place in the philosophy of nature" (*Essays on the Active Powers of the Human Mind* [*EAPHM*], 1788, *Works*, II, 526). It is significant that whereas Newton's disclaimer of hypothesis was confined to "experimental philosophy", Reid broadens it to apply to all of the philosophy of nature.

39 Reid's contemporary, James Gregory, perceived that the then predominant disdain for hypotheses was due to such terminological ambiguities as I have described briefly above: "The prejudice against hypotheses which many people entertain is founded on the equivocal signification of a word. It [viz., 'hypothesis'] is commonly confounded with theory . . . " (Quoted by Dugald Stewart in his *Works* [London, 1854], II, 300).

40 It would scarcely be an exaggeration to say that a great deal of the history of 18th-century methodological thought could be gleaned from carefully attending to the changing nuances in the meaning of this crucial term.

41 Reid was continually claiming that Newton was carrying on Bacon's work. In his

Intellectual Powers (p. 436) he asserted that "Lord Bacon first delineated the only solid foundation on which natural philosophy can be built; and Sir Isaac Newton reduced the principles laid down by Bacon into three or four axioms which he calls *regulae philosophandi*." In his *Brief Account of Aristotle's Logic* (*Works*, II, 712), he insisted that Newton "in the third book of his 'Principia' and in his 'Optics', had the rules of the *Novum Organum* constantly in his eye". Or, elsewhere, " . . . the best models of inductive reasoning that have yet appeared, which I take to be the third book of the *Principia* and the *Optics* of Newton, were drawn from Bacon's rules" (*Human Mind*, p. 200).

42 See (5) in the previous section of this chapter.

43 It was a common 18th-century mistake to write about Newton as if he invariably kept the *Novum Organum* at his finger-tips. For an early and influential example of this tendency, see the Introduction to Pemberton's *View of Sir Isaac Newton's Philosophy*. It is perhaps significant that among Reid's surviving manuscripts is a *Precis* of Pemberton's *View*.

44 "The true method of philosophizing in this: From real facts, ascertained by observation and experiments, to collect by just induction the laws of Nature, and to apply the laws so discovered, to account for the phenomena of Nature" (*Intellectual Powers*, pp. 271 ff.).

45 "The whole object of natural philosophy, as Newton expressly teaches, is reducible to these two heads: first, by just induction from experiment and observation, to discover the laws of nature; and then, to apply those laws to the solution of the phaenomena of nature. This was all this great philosopher attempted, and all that he thought attainable" (*Active Powers*, p. 529).

46 Reid to Kames, 16 December 1780, *Works*, I, 57; cf. p. 59.

47 Even Dugald Stewart, one of Reid's most enthusiastic admirers, admits that Reid was less explicit than he might have been about the sort of inductions he endorsed. In an account of Reid's life written in 1803, Stewart noted that: "it were perhaps to be wished that he [Reid] had taken a little more pains to illustrate the fundamental rules of that [inductive] logic the value of which he estimated so highly . . . " (in Reid, *Works*, I, 11). For an interesting contemporary critique of Reid's theory of induction, see Joseph Priestley's *Examination of Dr. Reid's Enquiry* (London, 1774), pp. 110 ff.

48 *Human Mind*, p. 163. Elsewhere he asserts that "all our curious theories . . . so far as they go beyond a just induction from facts, are vanity and folly . . . " (ibid., pp. 97–98). Reid sees the "slow and patient method of induction" as "the only way to obtain any knowledge of nature's work" (*Intellectual Powers*, p. 472). Again, he suggests that the discovery of *inductive* laws "is all that true philosophy aims at, and all that it can ever reach" (*Human Mind*, p. 157).

49 *Human Mind*, p. 202. Reid was convinced that Bacon was the unquestioned founder of inductive logic. In his *Brief Account of Aristotle's Logic*, he wrote: "after men had laboured in the search of truth near two thousand years by the help of syllogisms, Lord Bacon proposed the method of induction, as a more effectual engine for that purpose. His *Novum Organum* ought therefore to be held as a most important addition to the ancient logic. Those who understand it . . . will learn to hold in due contempt all hypotheses and theories" (Pp. 711–12).

50 "Since Sir Isaac Newton laid down the rules of philosophizing, in our inquiries into

Nature, many philosophers have deviated from them in practice; perhaps few paid that regard to them which they deserve" (*Intellectual Powers*, p. 251).

51 Birkwood Collection, box 2131.7, parcel 2. The full list of entries is as follows:

1. "Definition of body & explication ... of its primary qualities"; 2. "Laws of philosophizing from Sir Isaac Newton's Princ. Lib. 3"; 3. "Def. [initions] from same Lib. 1"; 4. "Three Laws of Nature ... "; 5. "Gravity ... "; 6. "Attraction of Cohesion"; 7. "Corpuscular attraction"; 8. "Magnetism"; 9. "Electricity." The copyright on this and all subsequent material quoted from the Birkwood Collection belongs to the Library of King's College, Aberdeen. The author is grateful to King's College for permission to quote from those papers.

52 Birkwood Collection, box 2131.5 (8), f. 37.

53 Ibid., box 2131.6 (II) (53).

54 In his unpublished "Lectures on Natural Philosophy" (1758), in the Library of the University of Aberdeen, Reid calls the *regulae* "the rules for reasoning by induction" (p. 7).

55 "Sir Isaac Newton, the greatest of natural philosophers, has given an example well worthy of imitation, by laying down the common principles or axioms, on which the reasonings in natural philosophy are built ... [the *Regulae*] are principles which, though they have not the same kind of evidence that mathematical axioms have; yet have such evidence that every man of common understanding readily assents to them ... " (*Intellectual Powers*, p. 231).

56 16 December, 1780, *Works*, I, 57. Compare this with Newton's observation that: "As in Mathematics, so in Natural Philosophy, the Investigation of difficult Things by the Method of Analysis, ought ever to precede the Method of Composition. This Analysis consists in making Experiments and Observations, and in drawing general Conclusions from them by Induction ... By this way of Analysis we may proceed from Compounds to Ingredients, and from Motions to the Forces producing them; and in general from Effects to their Causes, and from particular Causes to more general ones, till the Argument ends in the most general. This is the Method of Analysis: And the Synthesis consists in assuming the Causes discover'd and establish'd as Principles, and by them explaining the Phaenomena proceeding from them, and proving the Explanations" (*Opticks*, ed. Cohen [New York, 1952], pp. 404–5).

57 Cf. Newton's Regula IV where he concedes that "propositions inferred by general induction from phenomena ... may either be made more accurate, or liable to exceptions [i.e. refuted]".

58 "There must be many accidental conjunctions of things as well as natural connections; and the former are apt to be mistaken for the latter ... Philosophers, and men of science, are not exempted from such mistakes" (*Human Mind*, p. 197). The evidence that scientific laws "have no exceptions, as well as the evidence that they will be the same in time to come as they have been in time past, can never be demonstrative" (*Intellectual Powers*, p. 484). Cf. ibid., p. 272.

59 The second rule was this: "And so to natural effects of the same kind are assigned the same causes, as far as they can be."

60 Rule III began as follows: "The qualities of bodies which admit neither intension nor remission, and which belong to all bodies on which one can make experiments, are to be taken as the qualities of all bodies whatsoever."

[61] *Human Mind*, p. 199. This axiom is an exact translation of Newton's remark in the second *regula* to the effect that "effectuum naturalium ejusdem generis easdem esse causas." Elsewhere, Reid remarks that the second rule "has the most genuine marks of a first principle" (*Intellectual Powers*, p. 451).

[62] "Take away the light of this inductive principle, and experience is as blind as a mole; she may, indeed, feel what is present, and what immediately touches her; but she sees nothing that is either before or behind, upon the right hand or upon the left, future or past" (*Human Mind*, p. 200).

[63] "A natural philosopher can prove nothing, unless it is taken for granted that the course of nature is steady and uniform" (ibid., p. 130; cf. ibid., p. 198).

CHAPTER 8

THE EPISTEMOLOGY OF LIGHT: SOME METHODOLOGICAL ISSUES IN THE SUBTLE FLUIDS DEBATE

INTRODUCTION

There is wide agreement that new scientific theories often provoke a protracted discussion of their epistemic merits and their metaphysical presuppositions. Action-at-a-distance, vital forces, evolutionary theory, and the atomic debates are familiar cases in point. It has not, however, been widely appreciated that subtle fluid theories during their hey-day in the 18th and 19th centuries produced a philosophical discussion which was at least as deep as, and probably more far-reaching than, those associated with atomism and non-contact action. The aim of this chapter is to explore briefly some of the philosophical aspects of 'the ether debates', in order to document the impact of those debates on the fortunes of subtle fluid physics and on the 19th-century revision of the methodology of empiricism.

Although 'mediumistic' explanations have been a recurrent feature of science since antiquity, I want to discuss the state of the art from the mid-18th to the mid-19th centuries. It is chiefly this period that exhibits a very striking interaction between the physics of subtle fluids and empiricist epistemology, an interaction which will be the focus of this study. My claim will be that during this period the character of this interaction shifted profoundly in ways that were to modify both science and philosophy.

The central theses of this essay will be these:

(1) that the epistemology prevalent in the second half of the 18th century was altogether incompatible with the various ethereal theories which emerged in the natural philosophy of that period;
(2) that some of the early proponents of etherial explanations chose to abandon or modify that prevailing epistemology so as to provide a philosophical justification for theorizing about the ether;
(3) that the modifications so introduced were unconvincing and inadequate, leaving the scientific status of ether theories very unclear by the beginning of the 19th century;
(4) that the emergence of the optical ether in the early 19th century prompted a more radical critique of classical epistemology, a critique

that produced some innovative and historically influential methodological ideas.

The First Phase, 1740–1810

Our story should begin, as any account of Enlightenment epistemology must, by recalling the triumph of Newtonian mechanics and the trenchant inductivism associated with Newton's achievement. As numerous authors have shown, the half-century following publication of *Principia* was marked by a growing antipathy to hypotheses and speculations.[2] Induction and analogical reasoning were the rage and Newton's doctrine of *verae causae* – adumbrated in dozens of 18th-century glosses on his first *regula philosophandi* – was thought to exclude the postulation of any entity or process not strictly observable.[3] Whether we look to Berkeley and Hume in Britain, to 'sGravesande and Musschenbroek in the Netherlands, or to Condillac and D'Alembert in France, the refrain was similar: speculative systems and hypotheses are otiose; scientific theories must deal exclusively with entities which can be observed or measured. For half a century, many natural philosophers sought to develop theories satisfying those demanding strictures; 'moral philosophers' (e.g., Berkeley, Condillac, and Hume), for their part, explored the logical and epistemological ramifications of this new view of the nature of *scientia*.

However, long before epistemologists of science were able to digest these new challenges to the traditional demonstrative ideal of science, scientific developments themselves conspired to produce a significant shift. For, especially during the period from 1745 to 1770, many emerging theories within the sciences moved well beyond the inductive, observational bounds imposed by erstwhile empiricists (whether Newtonian or otherwise). Nowhere is this clearer than with respect to the development of mediumistic or ethereal explanations. In the 1740s alone, there were at least half a dozen major efforts to explain the behavior of observable bodies by postulating a variety of invisible (and otherwise imperceptible) elastic fluids. In 1745, Bryan Robinson published his *Sir Isaac Newton's account of the aether*. A year later Benjamin Wilson's *Essay towards an explication of the phenomena of electricity, deduced from the aether of Sir Isaac Newton* appeared. Of greater moment, Benjamin Franklin developed his account of electricity as a subtle fluid; the Swiss physicist George LeSage articulated an ethereal explanation of gravity and chemical combination; and the highly controversial David Hartley embarked on a program, culminating in his *Observations on man* (1749), to give a mechanistic theory of mind and perception, whose crucial ingredient is the transmission of vibrations in a subtle fluid or ether through the central nervous system. By the 1760s, the scientific literature abounded with ethereal explanations of heat, light, magnetism, and virtually every other physical process.

Two general points about these developments are especially relevant for our purposes. In the first place, by the 1770s, ethereal or subtle fluid explanations were very widespread among natural philosophers (with the exception, soon to be explained, of many Scottish scientists). Secondly, such explanations inevitably breached the prevailing epistemological and methodological strictures of the age; strictures which, as already noted, would not countenance the use of theoretical or 'inferred' entities to explain natural processes. (After all, an entity which is regarded as in principle unobservable is scarcely consistent with an empiricist epistemology which restricts legitimate knowledge to what can be directly observed.)

Indeed, there was scarcely any domain of scientific theorizing in the 18th century which left as much scope for speculative hypotheses about unseen agents as did ether theories. As Joseph Priestley remarked in his *History and present state of electricity*:

> Indeed, no other part of the whole compass of philosophy affords so fine a scene for ingenious speculation. Here the imagination may have full play, in conceiving of the manner in which an invisible agent produces an almost infinite variety of visible effects. As the agent is invisible, every philosopher is at liberty to make it whatever he pleases, and ascribe to it such properties and powers as are most convenient for his purpose.[4]

Not a tolerant epistemology at the best of times, classical empiricism (by which I mean the empiricism of thinkers as diverse as Berkeley, Condillac, Hume, and Reid) left no scope for entities like the ether. A few natural philosophers of the period failed to perceive the tension between the received epistemology and etherial theorizing. Leonhard Euler, for instance, could simultaneously maintain that the transmission of light depended upon vibrations in an imperceptible medium *and* insist, in his *Letters to a German Princess*, that science should proceed by enumerative induction, eschewing all non-observable entities. But most scientists and philosophers of the time saw the strain between the emergence of subtle fluid theorizing on the one hand and the subscription to a naive inductive empiricism on the other. Among this latter group, some – such as Thomas Reid – were persuaded that epistemological doctrines took priority over physical theories and thus should be allowed to legislate fluid theories out of the scientific arena. Others, like Hartley and LeSage, saw that option as self-defeating, and preferred instead to seek to develop new and more liberal versions of empiricist epistemology which could sanction subtle fluid theories. I want to examine both of these reactions.

The Partisans of the Ether. Among the most persistent, not to say the most notorious, proponents of subtle fluid theories in the last half of the 18th century were David Hartley and George LeSage. Hartley foresaw many explanatory roles for the subtle, elastic fluid which he called an 'aether'. Among them, explaining the transmission of heat, the production of gravity, electricity, and magnetism. Hartley's central concern, however, was to utilize the ether, or rather vibrations within the ether, to explain a large range of problems about perception, memory, habit and other activities of the mind. On Hartley's view (which was a detailed elaboration of one of the more speculative conjectures of Newton's *Opticks*),[5] the brain and nervous system are filled with a highly subtle fluid which transmits vibrations from one point in the perceptual system to another. The vibrations in the ether, which are initiated by some external stimuli, subsequently cause the medullary matter composing the nerves and brain to vibrate in ways that are characteristic of the stimuli in question. In his *Observations on Man* (1749), Hartley utilized the vibrations in the 'nervous' ether to explain a remarkably divergent range of phenomena, including 'sensible pleasure and pain' (pp. 34–44), sleep (45–55), the generation of simple and complex ideas (56–84), voluntary and involuntary muscular motions (85–114), the sensation of heat (118–25), ulcers (127), paralysis (132–4), taste (151–79), smell (180–90), sight (191–222), hearing (223–38), sexual desire (239–42), memory (374–82), and the passions (368–73). Indeed, most of the 500-odd pages of part one of the *Observations on Man* consist in a litany of phenomena which are explicable on Hartley's hypothesis of a vibratory ether.

As Hartley was perfectly aware, the whole structure of his argument was radically out of step with the prevailing inductivist temper of the age. Nowhere did Hartley attempt 'to deduce the ether from the phenomena'; nowhere could he use Baconian techniques of eliminative induction to establish the epistemic credentials of his enterprise. Neither could he point to any direct (i.e., non-inferential) evidence for the existence of his ubiquitous subtle fluid; hence it is no *vera causa*, as Thomas Reid was quick to point out. Rather, what Hartley had to settle for is a kind of *post hoc* confirmation. His arguments invariably have the structure of Peircean 'abductions':

Here is phenomenon x.
But if there were an ether, then x.
(Probably) there is an ether.

In short, straightforward hypothetico-deduction was the official methodology of the *Observations on Man*.

One need hardly add that therein lay the trouble. For however tolerant later generations were to be about *post hoc* confirmations and the abductive schema, Hartley's contemporaries – as he knew full well – viewed hypothetical reasoning as inherently fallacious (after all, 'abduction' *is* a form of the fallacy of affirming the consequent). Hartley's primary defense of his procedures involved a stress on *the wide range of confirming instances* which his theory could lay claim to. He suggested that its broad explanatory scope compensates for the unobservability of its explanatory agents and mitigates against its failure to exhibit a traditional inductive warrant. As he remarks early in the *Observations on Man*:

> Let us suppose the existence of the aether, with these its properties, to be destitute of all direct evidence, still, *if it serves to explain a great variety of phenomena, it will have an indirect evidence in its favour by this means.*[6]

Hartley likened the search for deep-structured theories like his own to the process of decoding a message. Just as the decypherer's task is to find a code which will render the encoded message intelligible, the natural philosopher's job is to find some hypothesis which will save the phenomena. The latter, like the former, must be content with *indirect* evidence:

> the decypherer judges himself to approach to the true key, in proportion as he advances in the explanation of the cypher; and this without any direct evidence at all.[7]

But Hartley must have known this analogy, baldly stated, would not take him very far. If he was to establish that it was scientifically respectable to speak about unobservable entities for which there could be no direct evidence, then he would have to re-orient the epistemological convictions of an age which took the view that, where hypotheses were concerned, indirect evidence was no evidence at all.

To that end, he composed a long section of the *Observations* dealing with "Propositions and the Nature of [rational] Assent". The unambiguous aim of that methodological excursus is to show that enumerative and eliminative induction are not the only routes to knowledge. He began the section on a then familiar note, to wit, that the methods of induction and analogy are the soundest methods of inquiry in natural philosophy. Indeed, he went so far as to give an associationist and vibrationist account of the mechanisms whereby repetitions of particular instances of a generalization reinforce one another so as to habituate us to accept the generalization of which they are instances.

Apart from the relatively novel assimilation of the method of induction to the calculus of probabilities, Hartley was here covering ground which was familiar territory to his contemporaries.

But Hartley went on to insist that the techniques of induction and analogy do not exhaust the methodological repertoire of the natural philosopher. There is, in addition, the *method of hypothesis*. A propos the Newtonian insistence that 'hypotheses have no place in experimental philosophy', Hartley was uncompromising: "It is in vain", he explained, "to bid an enquirer form no hypothesis. Every phenomenon will suggest something of this kind." Those who pretend to make no hypotheses are deceiving themselves and confuse their speculative hypotheses with "genuine truths . . . from induction and analogy".[9] Since the mind willy-nilly forms hypotheses when confronted by any phenomenon, it is far better to acknowledge tentative hypotheses than to acquiesce unwittingly in them:

he that [explicitly] forms hypotheses from the first, and tries them by the facts, soon rejects the most unlikely ones; and, being freed from these, is better qualified for the examination of those that are probable.[10]

Moreover, Hartley insisted, the examination and testing of hypotheses, even false ones, has the heuristic advantage of leading us quickly to the discovery of new facts about the world which we would be otherwise unlikely to discover:

The frequent making of hypotheses, and arguing from them synthetically, according to the several variations and combinations of which they are capable, would suggest numerous phenomena, that otherwise escape notice, and lead to *experimenta crucis*, not only in respect of the hypothesis under consideration, but of many others.[11]

But even granting (as some of the most trenchant inductivists were willing to)[12] that hypotheses can be of heuristic value, Hartley was still confronted with the problem of explaining the circumstances under which it is legitimate, contra-inductivism, to accept or believe an hypothesis involving unobservable entities. Unless it could be shown that there are some circumstances in which an hypothesis – not generated by inductive methods – warrants acceptance, Hartley's own program for a hypothetical science of mind would be without foundation. Hartley himself propounds the conundrum:

But in the theories of chemistry, of manual arts and trades, of medicine, and, in general, of the powers and mutual actions of the small parts of matter, the uncertainties and perplexities are as great, as in any part of science. For the small parts of matter, with their actions, are too minute to be the objects of sight; and we are neither possessed of a detail of the phenomena sufficiently copious and regular, whereon to ground an [inductive] investigation; nor of a method of investigation subtle enough to arrive at the subtlety of nature . . . .[13]

It is disappointing that, after much fanfare, Hartley's defense for believing a speculative hypothesis that explains many phenomena took him no further than the early pages of *Observations on Man* had done. Invoking again the cypher analogy, Hartley merely insisted that if an hypothesis is compatible with all the available evidence, then that hypothesis "has all the same evidence in its favour, that it is possible the key of a cypher can have from its explaining that cypher".[14] In a nutshell, Hartley's method of hypothesis boils down to the claim that an hypothesis warrants belief if it has a large number of known positive instances and no known negative instances. *Confirmed explanatory scope* thus functioned for Hartley as the decisive criterion for the acceptability of hypotheses.

Hartley immediately conceded that this criterion does not guarantee that the hypotheses it licenses will be true or that they will even stand up to further testing. They will possess none of the reliability (then) associated with the methods of induction and analogy. But, given the inevitability of hypotheses, what (he seems to ask) is the alternative? As he puts it,

the best hypothesis which we can form, i.e., the hypothesis which is most conformable to all the phaenomena, will amount to no more than an uncertain conjecture; and yet still it ought to be preferred to all others, as being the best that we can form.[15]

Not surprisingly, this epistemology carried little weight with most of Hartley's inductivist contemporaries. As they could point out, there were many rival systems of natural philosophy which – after suitable *ad hoc* modifications – could be reconciled with all the known phenomena. The physics of Descartes, the physiology of Galen, and the astronomy of Ptolemy would all satisfy Hartley's criterion. There was, in Hartley's approach to the epistemology of science, nothing which would discredit the strategy of saving a discarded hypothesis by cosmetic surgery or artificial adjustments to it.[16] As his critics pointed out, the great Newtonian epistemological innovation involved the insistence that 'saving the phenomena' or merely explaining the known data was an insufficient warrant for accepting a theory. Neither Newton nor his followers would quarrel with the view that it was a *necessary* condition of the acceptability of a theory that it must fit all the available data;[17] but they would not brook Hartley's transformation of this plausible *necessary* condition of theoretical adequacy into a *sufficient* condition for theory acceptance. Moreover, it was quite clear to any perceptive reader of Hartley's epistemological writings, that they were meant to rationalize his ethereal neurophysiology. Accepting the former meant acquiescing in the latter, which few were willing to do.

If Hartley chose to take on the inductivists somewhat obliquely, conceding to them that induction and analogy were sound modes of inference, expecting (but not receiving) in return an admission that the method of hypothesis too had its use, a more direct frontal attack on the prevailing epistemology came from another partisan of the ether, Hartley's Swiss contemporary George LeSage. By his own account, LeSage discovered in 1747 the theory which was to make him alternatively acclaimed and notorious for well over a century. LeSage's approach involved postulating a medium surrounding all bodies. The corpuscles constituting this ether move in all directions and occasionally impact upon the particles constituting observable physical objects. The latter are 'semi-permeable' to streams of ethereal particles, i.e., most etherial particles will pass completely through a macroscopic object (chiefly because the volume of its constituent particles is always a very small proportion of the space occupied by the body). Some, however, will collide with particles of the body; when they do, there will be appropriate transfers of momentum, with the ethereal particles rebounding and with the atoms of the body moving in the reverse direction. This kinematic ether is utilized by LeSage to explain a wide diversity of phenomena. In an article in *Mercure de France*, he utilizes it to explain weight;[18] in his *Essai de chimie mècanique,*[19] it is invoked to explain many phenomena of chemical affinity; still more significantly, he uses this approach in 1764[20] and again in 1784[21] to develop his famous kinematical model of gravitation.

The details of LeSage's ether model need not detain us here.[22] It was sympathetically, if critically, evaluated by Maxwell a century ago and its mathematical and physical articulation was extensively explored by (among others) Preston, Kelvin, Croll, Farr, George Darwin, and Oliver Lodge about the same time. For our purposes, a very brief summary of LeSage's model should suffice. LeSage explains gravity by assuming that there are streams of ethereal corpuscles flowing into the world from every direction. As pointed out above, ordinary bodies are highly porous and thus most of these 'ultramondain' corpuscles move through a body with no interaction. A few, however, will impact with constituent corpuscles of the body. The result of such impacts is that some ethereal particles reverse their direction while the body itself has a net force exerted on it by the collisions. In a one-body universe, there would be no resultant motion since there would be an equal number of collisions on all sides of the body. But if we introduce a second large object into this universe, each will act as a partial shield against the 'ultramondain' corpuscles. This will lead to a pressure differential, with each body undergoing fewer collisions on one side than the other; as a result, each body will tend to move toward the other. In this way, the qualitative character of gravitational attraction is attained via contact action rather than action-at-a-distance. The quantitative features of gravity (namely, its relation to the square of the distance and to the masses of the bodies) are explained in LeSage's full-blown *Traité de physique mécanique* (published posthumously by Pierre Prevost).[23] Although many of his predecessors (e.g., Fatio, Daniell, Bernoulli) had sought mechanical explanations for gravitational attraction, LeSage's was the only one to emerge from the 18th century as a *prima facie* physically adequate model of gravitational interaction. As James Clerk Maxwell remarked of LeSage's theory:

> Here, then, seems to be a path leading towards an explanation of the law of gravitation, which, if it can be shown to be in other respects consistent with facts, may turn out to be a royal road into the very arcana of science.[24]

But that reasonably flattering pronouncement by Maxwell is a far cry from the almost universal reaction of LeSage's contemporaries to his ethereal models. Immediately upon its publication (and, since LeSage widely pre-circulated his ideas, in many cases before they were published), LeSage's theory was subjected to a steady stream of abuse.

Roger Boscovich called LeSage's system a "purely arbitrary hypothesis", for which there is no direct proof.[25] The French astronomer Bailly, protesting against LeSage's model because it postulated hidden or unobservable entities, insisted that "nous ne connoissons la nature que par son extérieur, nous ne pouvons la juger que par celles de ses lois qu'elle nous a manifestées . . . "[26] During the 1760s, Leonhard Euler wrote several encouraging letters to LeSage about the latter's work. Nonetheless, Euler made it clear that

> je sens encore une trés-grande repugnance pour vos corpuscles ultramondains, et j'aimerois toujours mieux d'avouer mon ignorance sur la cause de la gravité que de recourir à des hypothèses si etranges.[27]

As these few passages suggest, the reaction to LeSage's model was not only largely negative; *the grounds for criticism were generally epistemological rather than substantive*. LeSage's critics were claiming that hypotheses about unobservable entities were no part of legitimate science. As early as 1755, LeSage complained that the common objection to his system of ultramondain corpuscles was that "mon explication ne peut être qu'une hypothèse".[28] In 1770, he worried to a correspondent that unless his system is presented very carefully, "on le jugeroit sur l'etiquette comme une de ces hypothèses en l'air".[29]

By 1772, LeSage became convinced that his theory was not getting the hearing it deserved because no one would evaluate it on its scientific merits. Instead, they dismissed it as a mere hypothesis. He claimed it to be an "almost universal prejudice" of his age that any theory that deals with unobservable entities cannot be regarded as genuinely scientific.[30] A decade later, LeSage had become so convinced that he could not get a fair hearing for his views that he withdrew publication of what was to be his *magnum opus*. As he bitterly explains:

> Puisque vos physiciens sont si prévenus contre le possibilité d'établir solidment l'existence de mes agens imperceptibles, très-propres cependant à rendre intelligibles les attractions, affinités, et expansibilités, que constituent à present toute la physique, je suspendrai encore quelque temps la publication des ouvrages que je préparois sur ses agêns ....[31]

If the story ended here, it would amount to just one more case of a scientist whose works were suppressed because they were out of tune with prevailing epistemological fashion. But the LeSage case is more interesting and more important than that, because the difficulties LeSage encountered with his physics prompted him to respond in kind, i.e., to articulate a rival methodology to the dominant inductivist one. LeSage did this in a number of philosophical essays, which were designed to show that hypotheses in general, and ones involving imperceptible entities in particular, may have a sound epistemic rationale. LeSage dealt with this matter at great length in a much-quoted essay published posthumously by Prevost. Originally but unsuccessfully intended for publication as an article in the great *Encyclopèdie*, the essay was titled "On the method of hypothesis".[32] In it, LeSage sought to show that enumerative induction and the method of analogy – the two dominant methods advocated by the ubiquitous Newtonians – were not as fool-proof as their partisans claimed and that the method of hypothesis[33] was not so weak as its critics insisted.

As LeSage shrewdly recognized, the core presupposition of the method of enumerative induction is that a clear distinction can be made between what is observable and what is not. The traditional contrast between inductive methods and speculative ones was that the former stayed very close to sensory experience, whereas the latter moved a long way from it, thereby acquiring a much greater degree of uncertainty. LeSage would admit that there possibly is a distinction to be made between what is observable and what is not, but he insisted that rigid and exclusive adherence to the former would produce a very emaciated science. "Those", he said, "who disparage the method of hypothesis do not allow us to make conjectures, *except those which follow naturally and immediately from experience*."[34] Although this view is "repeated superstitiously", there is no "precise idea" which can be attached to it. For "what on earth is an immediate consequence deduced from the observation of a fact? The existence of that fact and nothing more."[35] But perhaps what was meant is that claims about the world which are closer to observation are better supported than those which are, as it were, several inferential steps away from sensory particulars. LeSage will have none of this. If, he says, the claims we make "are hasty, what is gained if they are also immediate?" And if the claims we make are well-evidenced, "what does it matter whether [they are] immediate or as far removed from the phenomena as the last propositions of Euclid are from his axioms?"[36]

LeSage's point is that any form of theorizing goes well beyond the available data, so that there is no viable distinction to be made between theories which do, and those which do not, go beyond the evidence. Given that we must extrapolate from the known to the unknown, and that such a process is usually a risky business, what methodological rules can we use to insure that such inferences are justified? The bulk of LeSage's essay is devoted to that task, and specifically to showing that the method of hypothesis is of greater moment in scientific reasoning than the rival methods of induction and analogy. As LeSage uses the term, 'the method of hypothesis' refers to any procedure which involves a comparison of the logical consequences of a theory with observations. The highest form of proof for an hypothesis would involve showing that all its consequences were true, that it exhibits "exact correspondence with the phenomena".[37] But, as he acknowledged, we are rarely if ever in a position to obtain such exhaustive evidence. Failing that,

> if the assumed cause [i.e., the hypothesis] is able to produce *all the presently known features* related to the principal effect, then it will have the highest degree of certitude that we can at present hope for in the circumstances.[38]

LeSage here added an important qualification. Before we accept an hypothesis which can explain all the available evidence, we must be sure that our evidence represents a large sample. Hypotheses which work well when the data are limited frequently break down as the data base is extended. LeSage insisted that our belief in a hypothesis should be a matter of degree: "the greater the number of facts with which [the hypothesis] agrees, the more faith we should have in it".[39] Pressing his analysis more closely, he stressed that mere agreement (i.e., logical consistency) between an hypothesis and the evidence is a very weak relation. If an hypothesis is to be solidly confirmed by a piece of evidence then it must entail that evidential statement and none of its contraries. The greater the specificity with which the hypothesis correctly entails what we observe, the greater it will be confirmed by those observations.[40]

The fact that hypotheses are not *generated* by an analysis of the evidence had been, for many of LeSage's contemporaries, a major strike against them. Inductivists, in particular, had stressed that the only legitimate theories are those which arise as generalizations from experience. LeSage replied that so long as we subject our hypotheses to a rigorous process of 'vérification', it does not matter how they were generated initially. LeSage was as contemptuous as the inductivists of hypotheses which are not, or cannot be, verified. But he urged that we ought not confuse the horrors of unverifiable hypotheses with the merits of verified ones. "Thousands of times it has been said: the abuse of something, however universal, must never be taken as an argument against its legitimate usage."[41] He maintained that Newton, in rejecting the method of hypothesis, "never realized that one could utilize a method of research whose pitfalls he had recognized so well!"[42]

LeSage's next ploy consisted in a lengthy demonstration that, Newton's '*hypotheses non fingo*' notwithstanding, Newton's work in optics and mechanics was permeated by hypotheses and hypothetical reasoning.[43] Similarly, the works of Kepler, Copernicus, and Huygens rested on hypothetical modes of inference.[44]

LeSage conceded that some hypotheses are spurious, singling out in particular the vortex theories of Descartes and his followers. "The seventeenth century", he sagely observed, "preferred to acknowledge every hypothesis, however implausible, while our century finds it more convenient to reject them all."[45] Noting that epistemology "too, is subject to the rule of fashion and prejudice", LeSage maintained that there is a coherent middle way, which allows the use of hypotheses so long as strong empirical constraints are put on them. LeSage concluded his essay by pointing out the limited scope of analogical inference, insisting that it is "inconclusive", "arbitrary," and "impractical" and, in most cases, parasitic upon the method of hypothesis.[46]

Throughout this essay, as well as LeSage's other methodological writings, there are two levels of motivation. At one level, LeSage was genuinely concerned about epistemological issues in the abstract and felt it was important, as a philosopher, to get as clear as he could about the logic of science. But, lurking in the background (as with Hartley's discussion) is LeSage the ether theorist, struggling desperately to get a fair hearing for a scientific theory which is being dismissed on (what LeSage regards as) flimsy methodological grounds. There is nothing untoward in all of this. We would expect, after all, that knowledge and the theory of knowledge are closely intertwined. Nor are the excursions of LeSage and Hartley into epistemology merely self-serving apologiae. They are, that, of course; but they are substantially more than that as well. Chiefly, they are well-reasoned attempts to articulate and defend a hypothetico-deductive methodology in the face of inductivist criticisms.

There is another methodological theme common to the work of Hartley and LeSage which sets them apart from most of their contemporaries: a commitment to the *progressive* character of science. Throughout the earlier history of epistemology, the prevailing view was that putative scientific theories or doctrines were to be judged as true or false *simpliciter* and that the authentic methodology of science should be one which would produce true theories more or less immediately; provided, of course, that appropriate rules of inquiry were obeyed. Hartley and LeSage both protested against this all-or-nothing view of scientific theories. Acknowledging that their own theories may be false (because they are at best only probable), they stressed the approximative character of scientific inquiry. They insisted that theories, once promulgated, could be amended and improved, bringing them into ever closer agreement with their objects. To illustrate the point, both likened the development of scientific theories to certain mathematical methods of approximation (in fact, both singled out the rule of false position and the Newtonian technique for approximating roots of equations as examples). As Hartley remarked,

> Here a first position is obtained, which, though not accurate, approaches, however, to the truth. From this, applied to the equation, a second position is deduced, which approaches nearer to the truth than the first . . . Now this is indeed the way, in which all advances in science are carried on . . . .[47]

LeSage took the process of long division as a paradigm case of successful approximation.[48] Each step in the process brings us another digit closer to finding the true quotient. The testing and correction of hypotheses against experiment is, LeSage maintained, a suitable parallel.

This ploy proved to be very useful for LeSage and Hartley. Confronted with the inductivists' insistence that the method of hypothesis is inconclusive, they could grant the point; confronted by the claim that their specific ethereal models might be false, they could concede the possibility. But as they saw it, neither the fallibility of the method of hypothesis nor the falsity of some theories produced by it need force one to the conclusion that hypothesis has no role to play in science. On the contrary, once one admits that science is approximative and self-corrective, it becomes possible to envisage a theory *which is both false and an important step forward*.

Ultimately, in fact, the epistemological views of these two thinkers turn out to be significantly more influential than the ether models they developed (as we shall see below). But before we turn to look at the later fortunes of both the method of hypothesis and the ether hypothesis, we need to examine the opposition more closely.

The Early Critics of the Ether and Its Epistemology. Although ethereal explanations of electricity, magnetism, gravity, and heat became increasingly common and respectable in the course of the second half of the 18th century, there were still many natural philosophers who refused to countenance them. This was particularly true in Scotland where, the Scottish 'enlightenment' notwithstanding, many of the major philosophers or scientists were unwilling to acknowledge the importance of such theories. If ethers were generally regarded by the Scots as unsavory, Hartley's nervous ether was reserved for special abuse. In the early decades of the 19th century, the Scottish philosopher, Thomas Brown noted, with some pride, that "it is chiefly in the southern part of the island that the hypothesis of Dr Hartley has met with followers".[49] What Brown says about Hartley's theories can be duplicated for almost all the other major ethereal doctrines of the late 18th century; they rarely made it past Hadrian's Wall. Few of the major natural philosophers of Scotland's enlightenment (including Cullen, Black, Leslie, Reid, and Hutton) embraced many of the imperceptible fluid hypotheses that were, by the 1770s, being widely discussed (and, in some cases, widely accepted) in England and throughout much of continental Europe.

The primary reason for opposition to ether theories was the widespread acceptance among Scotish philosophers and scientists of a trenchant inductivism and empiricism, according to which speculative hypotheses and imperceptible entities were inconsistent with the search for reliable science. This linking of opposition to ether theories with an inductivist philosophy is neatly summed up in the 'Aether' article for the first edition (1771) of the *Encyclopaedia Britannica* (which, in spite of its title, was a predominantly Scottish production). Barely was the ether defined ("the name of an imaginary fluid, supposed by several authors ... to be the cause ... of every phenomenon in nature"),[50] before the author of the article launched into a virulent attack on the method of hypothesis and a spirited defence of the inductivism of Newton and Bacon:

> Before the method of philosophising by induction was known, the hypotheses of philosophers were wild, fanciful, ridiculous. They had recourse to aether, occult qualities, and other imaginary causes . . . .[51]

The article insisted that "the way of conjecture . . . will never lead any man to truth".[52] These passages from the *Britannica* reflect the thorough-going inductivism which influenced most Scottish writing on science in the period. Much of it sprang from the very influential work of Thomas Reid, leader of the so-called 'common sense philosophers'. As I have shown above, Reid's works are replete with abusive attacks on the method of hypothesis.[53]

What is important for our purposes is the *explicit* linkage between Reid's repudiation of Hartley's ethereal speculations and his attack upon the method which undergirds it. In his *Essays on the intellectual powers of man* (1785), Reid discussed Hartley's vibratory hypothesis at length. In the course of rejecting Hartley's "hypotheses concerning the nerves and brain",[54] Reid embarked on the lengthiest methodological discussion in his opus. Reid quickly perceived that Hartley's *Observations on Man* ran directly counter to the Newtonian inductivism which Reid himself espoused. He noted, significantly, that the epistemological chapter of the *Observation on Man* was written to justify the methodology utilized in Hartley's research: "Having first deviated from [Newton's] method in his practice, [Hartley] is brought at last to justify this deviation in theory."[55] Reid claimed that Hartley was the only author he knew who rejected the principles of Newton's inductivism and "Dr Hartley is the only author I have met with who reasons against them, and has taken pains to find out arguments in defense of the exploded method of hypothesis".[56] It seems natural to infer, then, that most of Reid's methodological tirades against hypotheses were directed chiefly at Hartley since he alone, in Reid's view, had criticized "the true method of philosophizing . . ."

That 'true method', so far as Reid was concerned, is some form of *enumerative* induction.[57] Reid nowhere spelled out the rules of his form of induction (as even his followers had to concede),[58] but some of its features are clear. As he wrote in the *Intellectual powers of man*:

> The true method of philosophizing is this: From real facts, ascertained by observation and experiments, to collect by just induction the laws of Nature, and to apply the laws so discovered, to account for the phenomena of Nature.[59]

The nub of the issue concerns the kinds of inductive generalizations which we can perform on 'observations and experiments'. Reid was adamant that two things must be insisted on at this stage:

(1) that any entities postulated in a putative law should really exist "and not be barely conjectured to exist without proof"; and (2) all the known deductive consequences of the law must be true.[60]

Condition (1), which Reid intended as a gloss on Newton's idea of *verae causae*, was extensively discussed, both in his published work and in his correspondence.[61] More often than not, Reid's first condition amounted to the rule that the scientist is allowed to postulate *only those entities which are observable*. Ethers and other imperceptible fluids are thus, by their very nature, disqualified from legitimate scientific status.

It is sorely tempting to side epistemologically with Hartley and LeSage against Reid. After all, the method of enumerative induction is *not* rich enough to build science; speculations and conjectures are inevitable; it does count in favor of a theory that it can explain a wide variety of phenomena, even if that theory postulates imperceptible entities. But to see the debate solely in these terms is misleading. In large part, the epistemological issue at stake between the inductivists and the hypotheticalists in the 18th century is simply this: does a confirming instance of a theory automatically count as a ground or reason for accepting or believing the theory? Both Hartley and LeSage insisted that any and every confirming instance provides evidence for the theory that entails it; accumulate enough such instances and the theory becomes credible. Thomas Reid, like such later philosophers of science as Peirce, the Bayesians, and Popper, maintained that 'mere' confirming instances are not enough, that some additional demand must be met before we can legitimately say that true deductive consequences of a theory count as evidence for that theory.

What motivates this conviction in Reid's case is a sound intuition that many unsavory theories have some true consequences. If we were to regard every theory with some true consequences as well established, then we could never judge one theory to have stronger evidence than another since "there never was an hypothesis invented by an ingenious man which has not this [kind of] evidence in its favor. The vortices of Descartes, the sylphs and gnomes of Mr Pope, serve to account for a great variety of phenomena."[62] Reid believed that a theory should not be regarded as confirmed or well established merely because it was sufficient to save the appearances. Of course, he did not quarrel with the view that legitimate theories must be compatible with the available evidence. But where Hartley and LeSage were willing to regard this as a *sufficient* condition for an acceptance theory, Reid views it as a *necessary* but not sufficient condition.[63]

But in a classic case of babies and bath-water, Reid's requirement went too far. His very narrow observational construal of the ground for warrantedly asserting the existence of a thing left Reid completely unable to give an account of the success of the many deep-structural theories of his time. By demanding too much, his epistemology was altogether unable to come to grips with the contemporary theoretical sciences. But if Reid and the inductivists demanded too much, Hartley and LeSage ran the risk of requiring too little. After all, many vacuous explanations – witness the classic '*virtus dormitivus*' in Molière – have true deductive consequences. Neither Hartley nor LeSage provided any plausible criterion for distinguishing vacuously true theories from legitimate scientific ones. In the absence of such a distinction, they had no grounds for claiming that their theories should be accepted in lieu of a multitude of equally well-confirmed but vacuous ones.

In sum, by the late 18th century, neither the inductivists nor the hypothetico-deductivists had yet constructed a plausible epistemology of science; equally, the fortunes of ether theories were still unsettled, precisely because it was unclear whether there was genuine evidence for them. All this was to change profoundly in the next half century. Philosophers of science would articulate new and more detailed criteria for determining which deductive consequences of a theory were genuinely confirmatory, and ether theories, at least certain ether theories, would pass these criteria with flying colors.

The Second Phase, 1820–1850

The early 19th-century debate about imperceptible fluids and the methods for establishing their existence focussed chiefly upon the wave theory of light (with its seemingly attendant commitment to a luminiferous ether). On the epistemological side, most of the interest centers around the emergence of a new methodological criterion for evaluating hypotheses. In brief, this criterion, which was nowhere prominent in the late 18th-century debates about the methodological credentials of subtle fluids, amounts to the claim that an hypothesis which successfully predicts future states of affairs (particularly if those states are 'surprising' ones), or which explains phenomena it was not specifically designed to explain, acquires thereby a legitimacy which hypotheses which merely explain what is already known generally do not possess. The major figures in this part of the story are Herschel, Whewell, and Mill. To put the matter concisely, John Herschel accepted the new criterion, saw that it provides a rationale for the wave theory of light, but did not recognize how that criterion threatened the traditional inductivist program to which he was otherwise committed. Whewell accepted the new criterion, and used it both to defend the wave theory of light and to attack traditional inductive procedures. Finally, John Stuart Mill, perceiving that the new criterion undermines induction, repudiated the former and, along with it, the vibratory theory of light. It is this set of interconnections with which I shall be concerned in this section.

The wave theory of light was revived chiefly, of course, by Thomas Young and Augustin Fresnel in the early years of the 19th century. Through the first three decades of that century, opinion as to the merits of the wave or the corpuscular theory of light was very divided among natural philosophers.[64] If the wave theory could provide a more convincing account of interference than the corpuscular theory, the latter seemed better able to explain problems of stellar aberration and double refraction. (Some participants in the ether debates, such as Comte, actually believed – mistakenly – that the two theories were observationally equivalent and that it was a matter of complete indifference which theory one utilized.)[65]

By the late 1820s, however, the balance of opinion was shifting perceptibly towards the undular theory. There were many factors responsible for this, but perhaps the most important among them was the Fresnel-Poisson experiment. The idea for the experiment arose in response to an essay written in 1816 by Fresnel on the nature of diffraction. In this paper, Fresnel elaborated the wave theory of light and applied it to the explanation of diffraction phenomena. Poisson, a confirmed corpuscularian and member of a panel refereeing Fresnel's paper, observed that, according to the analysis of light which Fresnel was using, it would follow that the center of the shadow of a circular disc would exhibit a bright spot. This predicted result was highly unlikely; it contradicted both the corpuscular theory *and* the scientist's intuitive sense of what was 'natural'. Indeed, the fact that the wave theory possessed this bizarre consequence was seen, *prior* to performing the experiment, as a kind of *reductio ad absurdum* of it. But when the appropriate tests were performed, the wave theory was vindicated by a concordance between what it predicted and the observed results.

There are two obvious methodological construals of the outcome of this experiment. On one interpretation, the experiment functions as a Baconian *experimentum crucis*, proving the falsity of the corpuscular theory and the truth of the wave theory. (Interestingly, although precisely this construal was given to the latter Foucault experiments on the speed of light in water and air, this was not the dominant interpretation of this earlier result.) On another interpretation, and this was widely adopted at the time, the disc experiment can be viewed as providing convincing evidence for the wave theory by virtue of its successful prediction of a surprizing (i.e., unexpected on the background knowledge) observational effect. The logic that undergirds the former (crucial experiment) interpretation is the familiar logic of eliminative induction. But what provides the epistemic rationale for the latter interpretation is a set of developments within the methodology of hypothesis evaluation.

As we have already seen, throughout the 18th century, proponents of the method of hypothesis pointed to *post hoc* (and, according to the critics of that method, presumptively *ad hoc*) explanations of known phenomena as the chief vehicle whereby an hypothesis proves its mettle. But none of the early proponents of the method of hypothesis could show what distinguished arbitrary and vacuous hypotheses from genuine and worthy ones, since both classes possessed large sets of *post hoc* confirming instances.

During the 1820s and 1830, proponents of the method of hypothesis articulated machinery for distinguishing 'artificial' hypotheses from legitimate ones. Specifically, they insisted that a proper hypothesis is one which not only explains what is already known, but which also can be successfully extended beyond the initial range of phenomena it was designed to explain. Particularly if the hypothesis can *predict* results which are unusual or surprising, the hypothesis thereby loses its artificiality and becomes a legitimate contender for rational belief. What is involved in this modification of the method of hypothesis (a modification especially prominent in the work of Herschel and Whewell) is *nothing less than a re-definition of what constitutes evidence*. In stressing that an hypothesis must establish its credentials by going beyond its initial data base, Herschel and Whewell defined a promising *via media* between the earlier extremes of Hartley and LeSage, on the one hand, and Reid and the other inductivists, on the other. By claiming that nothing can be arbitrary about an hypothesis which is successfully tested against a body of evidence independent of the circumstances which the hypothesis was invented to explain in the first place, they thus defused the charge of arbitrariness which was traditionally directed against the method of hypothesis.

Given the familiarity of this idea of 'independent tests' to a modern reader, it is easy to underestimate how significant a shift in the history of methodology it represents.[66] Part of its importance lies in the stress it puts on testing claims against the unknown rather than the known. Before the method of hypothesis could be plausibly viewed as anything more than the logical fallacy which it had always been considered to be, something new was called for, something which would separate serious and legitimate hypotheses from bogus or specious ones. What suited the bill, at least so far as the 1830s were concerned, was what I shall call *the requirement of independent or collateral support*. In brief, this requirement amounted to the demand that before an hypothesis was credible, it must explain (or predict) states of affairs significantly different from those which it was initially invented to explain. Evidence of this kind might come from one of two sources: either a surprising prediction of unknown effects or a successful explanation of phenomena that were already known but which did not serve as the original base for the formulation of the hypothesis.

This methodological requirement of independent support ought not be confused with the earlier empiricist requirement that theories must involve '*verae causae*'. That earlier demand had nothing to do with the capacity of a theory to predict surprising results; it insisted, rather that the entities postulated in a theory must be directly observable or directly 'inferred from the phenomena'. In short, the methodological tradition of *verae causae* had rested upon a rigid distinction between directly observable entities and not-directly observable ones, endorsing the former and eschewing the latter. By contrast, what I am calling the requirement of independent support is indifferent to the question whether theoretical entities are observable. It focusses, rather, on the *epistemic* features of the statements which can be deduced from a theory.

A theory like LeSage's gravorific ether, even if it had been entirely successful in its explanatory ambitions, did not seem to possess independent support in the sense defined. More to the point, no epistemologist in the 18th century would have been impressed if it had, for the notion of independent support in the sense under discussion here, is very much a product of the early-19th century. By invoking this requirement, as we shall see, proponents of the optical ether were able to argue – as their 18th-century ethereal precursors could not – that there was some very impressive evidence available for a luminiferous ether, evidence which went well beyond the ability of that hypothesis merely 'to save the phenomena'. The Fresnel circular disc experiment (a confirmed prediction) as well as the successful extension of the wave theory to polarization, di-polarization, and double refraction betokened a degree of collateral support for the optical ether which earlier ethereal doctrines did not exhibit.

This set of issues was stressed with particular emphasis by both John Herschel and William Whewell. Herschel insisted that we cannot reasonably expect a theory to be a reliable predictor in the future unless it has also been so in the past. Unless we have seen that a theory enables "us to extend our views beyond the circle of instances from which it is obtained", then "we cannot rely on it".[67] Before we accept any theory we must try "extending its application to cases not originally contemplated ... and pushing the application of [it] to extreme cases".[68] Although Herschel frequently enunciates this requirement, and sees it as a vehicle for protecting us from "the unrestrained exercise of imagination ... arbitrary principles ... [and] mere fanciful causes",[69] he does not discuss its rationale at any length.

What he does, however, is to invoke this requirement repeatedly to show that the hypothesis of the optical ether is a sound one. Herschel points out that Young's wave theory, originally developed to explain reflection, refraction, and interference, was eventually applied with much success to the explanation of polarization and double refraction.[70] He points out that not only does (Fresnel's version of) the wave theory explain "perhaps the greatest variety of facts that have ever yet been arranged under one general head"[71] but adds that Fresnel's theory also predicted "a *fact* which had never been observed ... and all opinion was against it".[72] The confirmation of this surprising prediction did much, in Herschel's view, to establish the 'probability' of the undular hypothesis.

Like Herschel, Whewell finds it necessary to supplement the weak demands of the traditional method of hypothesis by further constraints. He elaborates these in a lengthy section of the *Philosophy of the Inductive Sciences* devoted to 'Tests of Hypotheses'. With LeSage and Hartley, Whewell insists on the minimal condition: "the hypotheses which we accept ought to explain phenomena we have observed".[73] More precisely, he stipulates that every hypothesis must be "consistent with *all* the observed facts".[74] But unlike Hartley and LeSage, who saw in such a requirement a *sufficient* condition for adequacy, Whewell argues that hypotheses "ought to do more than this: our hypotheses ought [successfully] to *fortel* [sic] phenomena which have not yet been observed".[75] At a minimum, these predictions should be borne out by testing the hypothesis against the phenomena "of the same kind as those which the hypothesis was invented to explain".[76]

The rationale for this rather more exacting requirement is precisely that it dissipates the air of arbitrariness that surrounds an hypothesis whose only known instances were those used in its generation:

> Men cannot help believing that the laws laid down by discoverers must be in a great measure identical with the real laws of nature, when the discoverers thus determine effects beforehand in the same manner in which nature herself determines them when the occasion occurs.[77]

Successful prediction of effects similar to those already known does much to increase our confidence in an hypothesis. "But the evidence in favour of our induction is of a much higher and more forcible character when it enables us to explain and determine [i.e., predict] cases of a kind different from those which were contemplated in the formation of our hypothesis."[78] Whewell's technical term for this particular mode of evidencing (which involves testing a hypothesis against types of processes or events different *in kind* from those it was devised to explain) is the '*consilience of inductions*'. In his view, this is the most impressive type of evidence which theories can possess. Whewell's stress on the special confirmatory value of successful predictions, and the contrast it marks with the earlier 18th-century discussions of the method of hypothesis, comes out very clearly in his unusual utilization of the decyphering analogy. We saw this analogy already in Hartley (and it occurs before him in Descartes and Boyle *inter alia*). In its pre-Whewellian form, the analogy had suggested that if a certain hypothetical assignment of letters to an encoded cypher produces an intelligible message, then this constitutes evidence that the hypothetical assignment is correct. In this version of the analogy, it is assumed that the entire cypher is known in advance to the decoder. In Whewell's version of the analogy, however, a portion of the cypher is initially concealed from the decoder and the test of his decoding is *whether he can predict* the character of the concealed cypher.[79] As Whewell's variation on this traditional analogy makes clear, his concern is to assign *differential weights to the confirming instances of a theory*.

Barely had Whewell defined this noion of consilience before he cited the wave theory of light *in extenso* as one of the few theories to have passed this demanding test. In quick succession, he ticked off the explanatory successes of the wave theory: reflection, refraction, colors of thin plates, polarization, double refraction, dipolarization, and circular polarization. By contrast, the emission theory exhibits "what we may consider the natural course of things in the career of a false theory".[80] It can well enough explain "the phenomena which it was at first contrived to meet; but every new class of facts requires a new supposition . . . as observation goes on, these incoherent appendages accumulate, till they overwhelm and upset the original framework."[81]

So impressed was Whewell by the predictive successes of the wave theory that it functions as one of his two paradigm cases of exemplary theory development (Newtonian gravitational theory being the other), and his elaborate 'Inductive Table of Optics' culminates in the wave theory. Even in his earlier *History of the Inductive Sciences*, Whewell saw the wave theory as the optical equivalent of Newtonian mechanics. After a lengthy discussion of what he then called 'the undulatory theory', he remarked:

We have been desirous of showing that the *type* of this progress in the histories of the two great sciences, Physical Astronomy and Physical Optics, is the same. In both we have many *Laws of Phenomena* detected and accumulated by astute and inventive men; we have *Preludial* guesses which touch the true theory . . . finally, we have the *Epoch* when this true theory . . . is recommended by its fully explaining what it was first meant to explain, and confirmed by its explaining what it was *not* meant to explain.[82]

This passage neatly foreshadows Whewell's philosophy of science (with its emphasis on the independent support requirement) and the key role which the wave theory played in its formulation.

But what of the opposition? As we have seen, both Herschel and Whewell based their endorsement of the wave theory on its satisfaction of an innovative and highly controversial methodological demand (i.e., the requirement of independent support). If my account of the connection between views toward optics and this methodological rule are correct, we should expect that those who did not accept the Herschel/Whewell methodology would not share their enthusiasm for the undular theory. Confirmation for this expectation is ready at hand in the works of John Stuart Mill. Mill was, of course, the arch-foe of Whewell during the 1840s and 1850s. Their respective philosophies of science exhibited divergences at almost every major point. Not the least of these differences was the disagreement of the two men about the relative confirmational value of different instances of a theory. As we have seen, Whewell maintained that a theory was better confirmed by predictive instances or by explaining phenomena it was not originally devised to explain than it was by instances which it was devised to explain. By contrast, Mill insisted that predictive successes, every bit as much as *post hoc* explanatory ones, are highly inconclusive; both, if taken as grounds for asserting the theories which achieve them, are highly fallacious.

Significantly, this controversy emerges in Mill's *System of Logic* with specific reference to the luminiferous ether. Mill's central point was that no number of confirming instances of a theory can establish it conclusively, chiefly "because we cannot have, in the case of such an hypothesis (viz., the optical ether), the assurance that if the hypothesis be false it must lead to results at variance with true facts".[83] The fact that the wave theory of light "accounts for all the *known* phenomena" does not warrant the view that it is "probably true".[84] He then turned to consider Whewell's (and Herschel's) claim that it is chiefly the predictive successes of the wave theory which render it likely. "It seems to be thought", Mill observed, "that an hypothesis of the sort in question is entitled to a more favourable reception, if, besides accounting for all the facts previously known, it has led to the anticipation and prediction of others which experience afterwards verified."[85]

Mill was scathing in his insistence that such a view flagrantly confounds the psychology of surprise with the methodology of support. Referring specifically to one of the predictions of the wave theory, he said:

> Such predictions and their fulfillment are, indeed, well calculated to impress the uninformed,[86] whose faith in science rests solely on similar coincidences between its prophecies and what comes to pass. But it is strange that any considerable stress should be laid upon such a coincidence by persons of scientific attainments. If the laws of the propagation of light accord with those of the vibrations of an elastic fluid in as many respects as is necessary to make the hypothesis afford a correct expression of all or most of the phenomena known at the time, it is nothing strange that they should accord with each other in one respect more.[87]

Mill was not taking exception to Whewell's *psychological* observation that many people are impressed by a theory that successfully makes surprising predictions. What he was calling for is a logical or epistemological account of why we should regard such instances as being of a privileged, probative character. Like Popper a century later, Whewell did not ultimately meet this challenge; he never showed why a theory's novel predictions should count for so much more, in terms of its epistemic appraisal, than its successes at explaining what it was devised to explain.

Nonetheless, Whewell tried to restate his case in his *Of induction, with especial reference to Mr. J. S. Mill's System of Logic* (1849). Whewell rightly summarized Mill's attack on him by saying that it amounted to the traditional inductivist charge that one ought not allow "hypotheses to be established, merely in virtue of the accordance of their results with the phenomena".[88] Whewell reiterated that his was not merely the old-fashioned method of hypothesis, for he adds the demands of independent support and consilience. As for Mill's charge that successful predictions should impress only the "ignorant vulgar", Whewell insisted that "most scientific thinkers . . . have allowed the coincidence of results predicted by theory with fact afterwards observed, to produce the strongest effects upon their conviction".[89] (In the same passage he refers to "the curiously felicitous proofs of the undulatory theory of light".) But most of Whewell's reply consists chiefly of pious hand-waving rather than cogent arguments; he never satisfactorily meets Mill's challenge to produce a plausible epistemological rationale for the requirement of independent support.

At another level, however, Mill had perhaps missed the point. It is one thing to stress, as he rightly did contra-Whewell, that one or two surprising, but confirmed, predictions do not prove the theory which produces them. But if I am right, the motivation for introducing the independent requirement was not to transform the method of hypothesis into a proof technique. Rather, the concern had been to find some way of reducing the arbitrariness and adhocness of hypotheses. So long as the only confirming instances which an hypothesis could claim were those used in its generation, there was no reason to expect that applications of it to further instances would be successful. After all, the hypothesis might simply have fastened on some non-causal or non-nomic accidents of the cases so far surveyed, and generalized these into a universal theory. But, as Herschel and Whewell observed, if the hypothesis could be successfully extended to cases or even to domains which were not used in its development, then it can no longer be claimed that the reliability of the hypothesis is limited to the phenomena which it was devised to explain.[90]

In suggesting that the requirement of independent support played a major role in the acceptance of the wave theory of light in the 1830s and 1840s, I do not mean to imply that all the proponents of the wave theory invoked this requirement. Indeed, many did not. As Geoffrey Cantor has shown in his valuable study of the early reception of the wave theory, many natural philosophers accepted that theory simply because it satisfied the older (Hartley/LeSage) condition of large explanatory scope.[91] Where Cantor has stressed the invocation of traditional and already well-established methodological doctrines (e.g., scope) for appraising the wave theory, my concern has been to show the extent to which the wave theory occasioned some major thinkers to break with the methodological orthodoxy and to articulate new tests for hypotheses.

CONCLUSION

By the 1850s, the wave theory of light (and the associated hypothesis of a luminiferous ether) enjoyed a degree of acceptance among natural philosophers which none of the subtle fluid ethers of the 18th-century had possessed. I have tried to show in this chapter that the major reason for the different receptions afforded to the two sets of theories is to be explained primarily in terms of the possession by the wave theory of certain epistemic or methodological features which the earlier ethers did not possess. But the moral of the tale is not to be found entirely in this difference; for even if earlier ethers had been predictive and independently testable (in the sense indicated above), it is not clear that they would have been any more widely accepted than they were. What was needed was a shift not only in physics but in epistemology as well, so that independent support could be accepted as a decided epistemic virtue. That shift came (I have argued) in the 1830s, provoked in part by the wave theory itself, which served as an epistemological archetype for such philosophers as Herschel and Whewell. There is, of course, a circularity here. But, far from being of the vicious variety, it reflects the kind of mutual dependence between theory and *praxis* which has always characterized science and philosophy at their best.[92]

NOTES

1 For a detailed discussion of the rationale for such interactions, see my *Progress and its Problems* (London, 1977), ch. 2.

2 See, for instance, Chapter 7 above and I. Cohen, *Franklin and Newton* (Philadelphia, 1956), *passim*.

3 That the prevailing 18th-century interpretation of Newton's first rule involves the demand that all theories must restrict themselves to purely observable entities is slightly a matter of conjecture. What can be said with some confidence is that British and French glosses on Newton's *Regula* generally construe it in this way. (For a typical and very influential discussion of Newton's first rule, see Thomas Reid, *Philosophical works*, 6th ed. (Edinburgh, 1863), vol. I, pp. 57, 236, 261, 271–2.)

4 J. Priestley, *History and Present State of Electricity*, 3rd ed. (London, 1775), vol. II, p. 16.

5 See the sixteenth query to Newton's *Opticks*.

6 D. Hartley, *Observations of man, his frame, his study, and his expectations* (London, 1791), vol. I, p. 15. (The work first appeared in 1749.) A very similar methodological argument had been made two years earlier by Bryan Robinson: "This *Aether* being a very general material Cause, without any Objection appearing against it from the Phaenomena, no Doubt can be made of its Existence: For by how much the more general any cause is, by so much the stronger is the Reason for allowing its Existence. The Aether is a much more general Cause then our Air: And on that Account, the Evidence from the Phaenomena, is much stronger in Favour of the Existence of the *Aether*, than it is in Favour of the Existence of the Air" (*A Dissertation on the Aether of Sir Isaac Newton* (London, 1747), preface, n. p.).

7 Hartley, *Observations*, I. p. 16. This decyphering analogy has a long pre-history among earlier proponents of the method of hypothesis. It can be found, for instance, in Descartes (*Oeuvres*, ed. Adam and Tannery, Paris, 1897–1957, vol. x, p. 323) and Boyle (*Royal Society, Boyle Papers*, vol. ix, f. 63), among other 17th-century writers. As I shall show below, it continued to be used by methodologists for well over a century after Hartley.

8 Hartley, *Observations*, I, p. 346.

9 Ibid.

10 Ibid.

11 Ibid., p. 347.

[12] Many inductivists distinguished between what has subsequently been called the contexts of discovery and justification. They were quite prepared to grant that hypothetical methods were useful for the former, but insisted that they had no role in the latter. As Thomas Reid succinctly put it, "Let hypotheses . . . suggest experiments, or direct our inquiries; but let just induction alone govern our belief" (Reid, *Works*, I, p. 251).

[13] Hartley, *Observations*, I, p. 364.

[14] Ibid., p. 350.

[15] Ibid., p. 341.

[16] Indeed, Hartley even seems to endorse such a course of action. If, he says, our suppositions and hypotheses "do not answer in some tolerable measure [to the real phenomena, we ought] to reject them at once; or, if they do, to add, expunge, correct, and improve, till we have brought the hypothesis as near as we can to an agreement with nature" (*Observations*, I, p. 345).

[17] Indeed, Newton's first *regula philosophandi* insisted that theories must be "sufficient to explain the appearances".

[18] [G. LeSage], 'Lettre à un academicien', *Mercure de France*, Mai 1756.

[19] Published in Paris in 1758.

[20] G. L. LeSage, "Loi, qui comprend, malgré sa simplicité, toutes les attractions . . . ", *Le Journal des scavans*, Avril 1764, pp. 230–4. He wrote later papers on this same topic in *the Journal des beaux-arts* (Nov. 1772) and (Feb. 1773) and in the *Journal de physique* (Nov. 1773).

[21] G. LeSage, "Lucrèce newtonien", *Mêmoires de l'académie royale des sciences et belles-lettres de Berlin* (Berlin, 1784), pp. 1–28. (This essay was reprinted in Prevost, *Notice*, pp. 561–604).

[22] Readers seeking details about LeSage's gravitational aether should consult either LeSage, "Lucrèce", or S. Aronson, "The gravitational theory of Georges-Louis de Sage", *The natural philosopher* 3 (1964), 53–73.

[23] Cf. P. Prevost, *Deux traités de physique mécanique* (Geneva, 1818).

[24] W. Niven, ed., *Scientific papers of James Clerk Maxwell* (London, 1890), vol. II, p. 474.

[25] P. Prevost, *Notice de la vie et des écrits de George-Louis LeSage* (Geneva, 1805), p. 358.

[26] Prevost, *Notice*, p. 300.

[27] Even an enlightenment liberal, Euler added, "Mais j'accorde très volontiers cette liberté à d'autres". Euler to LeSage, 8 September 1765, University of Geneva Library; Ms. Suppl. 512, f. 314^{r}. (LeSage's transciption of Euler's letter is Ms. Fr. 2063, f. 141^{r}.)

[28] Prevost, *Notice*, pp. 464–5.

[29] Prevost, Ibid., 237.

[30] He spoke of "la prétendue impossibilité d'établir solidement un système, qui roule sur les objects essentiellement imperceptibles." (Prevost, *Notice*, p. 264).

[31] Prevost, *Notice*, p. 242.

[32] The full text of the essay was published in Pierre Prevost, *Essais de philosophie* (Génève, An XIII), Vol. II, 258 ff. The original can be found in the University of Geneva Library, ms. Fr. 2019(2). Because there are some (largely minor) discrepancies between the printed version and the original, all my quotations shall be from the latter. I shall give in parentheses references to the appropriate paragraph in Prevost's text.

33 By the term 'method of hypothesis', LeSage meant the view that science "is conducted by the method of trial and error . . . by gropings followed by verification, by hypotheses which are then confirmed by their agreement with the phenomena" (University of Geneva Library, Ms. Fr. 2019(2). (para 5)).
34 University of Geneva Library, Ms. Fr. 2019(2) (para. 26).
35 Ibid.
36 Ibid.
37 University of Geneva Library, Ms. fr. 2019(2). (para 7). Joseph Priestley would similarly insist that if a scientist can frame his theory so as really to suit all the facts, "then it has all the evidence of truth that the nature of things can admit" (Priestley, *History*, p. 16). For allusions to the very early history of this principle, see Chapter 6.
38 University of Geneva Library, Ms. fr. 2019(2) (para. 7) (my italics).
39 University of Geneva Library, Ms. fr. 2019(2) (para. 15).
40 Cf. Ibid. LeSage even seemed to think, as did Hartley, that with a sufficiently large number of confirming instances, the hypothesis becomes virtually certain. As he wrote in his *Principes généraux de la téléologie*: "Plus les phénomènes sont nombreux et plus la précision est grande; plus aussi ils jugent avec assurence qu'il ne sauroit y avoir d'autre hypothèse sur le même sujet, qui ait les mêmes advantages" (Prevost, *Notice*, pp. 529–30).
41 University of Geneva Library, Ms. fr. 2019(2) (para. 18).
42 University of Geneva Library, Ms. fr. 2019(2) (para. 19).
43 He claimed that "almost all that the first two books of his *Principia* [i.e., Newton's] contain . . . is nothing more than a collection" of "curious hypotheses" (University of Geneva Library, Ms. fr. 2019(2) (para. 20)).
44 University of Geneva Library, Ms. fr. 2019(2) (paras. 23–5).
45 University of Geneva Library, Ms. fr. 2019(2) (para. 29). Lest the wary reader may think LeSage was painting an exaggerated picture of the Newtonians' aversion to hypotheses, it is worth saying that his view is borne out by many of his contemporaries. Thus, the *Encyclopédie* observed that Newton "et sur-tout ses disciples" are very opposed to hypotheses, regarding them as "le poison de la raison et la peste de la philosophie" (article on 'Hypothèse'). Across the channel, the Newtonian Benjamin Martin noted in the 1750s, "The philosophers of the present Age hold [hypotheses] in vile Esteem, and will hardly admit the name in their Writings; they think that which depends upon bare Hypothesis and Conjecture, unworthy the name of Philosophy" (B. Martin, *Philosophical Grammar*, 7th ed. (London, 1769), p. 19).
46 University of Geneva Library, Ms. fr. 2019(2) (paras. 30–8).
47 Hartley, *Observations*, I, p. 349.
48 University of Geneva Library, Ms. fr. 2019(2) (para. 5).
49 T. Brown, *Lectures of the Philosophy of the Human Mind*, 20th ed. (London, 1860), p. 279.
50 *Encyclopaedia Britannica* (Edinburgh, 1771), vol. I, p. 31.
51 Ibid.
52 Ibid.
53 See Chapter 7.
54 T. Reid, *Philosophical works*, W. Hamilton, ed., 2v. (Hildesheim, 1967), pp. 248–53.
55 Ibid., p. 250.
56 Ibid., p. 251.

57 Like Newton, in whose footsteps he sought to follow, Reid rejected *eliminative* induction, chiefly on the ground that we cannot perform the exhaustive enumeration of possible hypotheses which that method requires. (See, for instance, Reid, *Philosophical Works*, I, p. 250.)

58 As Reid's successor Dugald Stewart observed: "it were perhaps to be wished that [Reid] had taken a little more pains to illustrate the fundamental rules of that [inductive] logic the value of which he estimated so highly ... " (Reid, *Philosophical works*, I, p. 11).

59 Reid, *Philosophical works*, I, p. 271.

60 Ibid., p. 250.

61 See especially the exchange of letters between Reid and Lord Kames in Reid, *Philosophical works.*

62 Reid, *Philosophical works*, I, p. 251.

63 That it is a necessary condition is made clear by Reid's 'second requirement' cited above.

64 I shall not discuss the first methodological debate which the wave theory provoked, namely, that between Young and Brougham. I skip over it for two reasons: (1) it has already been investigated at length by G. Cantor in his 'Henry Brougham and the Scottish methodological tradition', *Studies in History and Philosophy of Science* 2 (1971), 69–78; (2) it represents a more vituperative but less substantial replay of the earlier ether debates I have discussed, with Brougham playing Reid to Young's Hartley.

65 See especially A. Comte, *Cours de philosophie positive*. (Paris, 1924), II, pp. 331–52. A general discussion of Comte's philosophy of science can be found in my Chapter 9.

66 Indeed, given the ubiquity of the requirement – or analogs of it – in contemporary philosophy of science, it is remarkable that its pre-history has not yet been explored. In an intriguing but false surmise, Karl Popper once observed that "successful new prediction – of new effects – seems to be a late idea, for obvious [sic] reasons; perhaps it was first mentioned by some pragmatist ... " (*Conjectures and refutations*, London, 1965, p. 247).

67 For a useful discussion of related issues, see V. Kavaloski, 'The '*vera causa*' principle', Ph.D. thesis, University of Chicago, 1974.

68 J. F. W. Herschel, *A Preliminary Discourse on the Study of Natural Philosophy* (London, 1830), p. 167; see also pp. 172, 203.

69 Ibid., p. 190.

70 Ibid., 259 ff.

71 Ibid., p. 32.

72 Ibid., pp. 32–3.

73 W. Whewell, *The Philosophy of the Inductive Sciences, Founded Upon Their History*, 2v. (London, 1847), II, p. 62.

74 Ibid.

75 Ibid.

76 Ibid., pp. 62–3.

77 Ibid., p. 64.

78 Ibid., p. 65. For a fuller discussion of these issues, see Chapter 10.

79 "If I copy a long series of letters, of which the last half dozen are concealed, and if I guess these aright, as is found to be the case when they are afterwards uncovered, this

must be because I have made out the import of the inscription" (*Philosophy of discovery* (London, 1860), p. 274).

80 Whewell, *Philosophy*, II, p. 72.

81 Ibid.

82 W. Whewell, *History of the Inductive Sciences from the Earliest to the Present Time*, 3rd ed. (London, 1857), II, p. 370.

83 J. S. Mill, *System of Logic, Ratiocinative and Inductive*, 8th ed. (London, 1961), p. 328.

84 Ibid.

85 Ibid.

86 In early editions of the *System*, he referred here to the "ignorant vulgar" rather than the "uninformed".

87 Mill, *System*, pp. 328–9.

88 W. Whewell, *On the Philosophy of Discovery* (London, 1860), p. 270.

89 Ibid., p. 273.

90 The Herschel-Whewell requirement of independent support has shown up in a new guise in the work of E. Zahar, especially his "Why did Einstein's programme supersede Lorentz's?" in C. Howson, ed., *Method and Appraisal in the Physical Sciences* (Cambridge, 1976), pp. 211–76. Zahar is no more successful than Herschel and Whewell were in providing a philosophical justification for the requirement. (For a latter-day 'Millean' critique of recent work in this area, see Laudan, *Progress*, pp. 114–18.)

91 G. Cantor, [1976]. There are many other dimensions of the ether debates which deserve serious exploration. For instance, some natural philosophers regarded the mathematical formalisms of the wave theory as well established but refused to regard this as evidence for the existence of a luminiferous ether. What was at stake was whether a theory could be 'accepted' without taking its ontological presuppositions seriously.

92 I have argued for the symbiotic character of the general relationship between science and philosophy in Laudan, *Progress*, chap. 2, and in Chapter 2, above.

CHAPTER 9

TOWARDS A REASSESSMENT OF COMTE'S 'MÉTHODE POSITIVE'

Judged by almost any criteria, Auguste Comte's theory of positivism was an influential doctrine in the history of the philosophy of science. His contemporaries took it very seriously indeed, whether they were his followers, like (the early) Mill[1] and Littré,[2] or his opponents, like Whewell.[3] His 20th-century successors, too, evidently attached some importance to Comte, since the phrase '*logical* positivism' cannot have been an entirely capricious choice of label by the philosophers of the Vienna Circle.[4] But in spite of the frequency with which the term 'positivism' is used and notwithstanding the fact that Comte is invariably cited as one of the important precursors of the Vienna Circle, remarkably little has been written about the details of Comte's theory of scientific method and his philosophy of science.[5] Apart from his celebrated theory of the three stages of intellectual history, which has been discussed at length, there exists as yet nothing like a detailed exegesis of Comte's views on issues like induction, prediction, hypotheses, and explanation. The tacit assumption appears to be that although Comte's general approach was interesting, influential and provocative, it would probably be both unrewarding and tedious (when one recalls the prolixity of Comte's prose) to push very far into an exploration of his views. However, it seems to me that such an inquiry is worth undertaking, not only because in its absence any claims about Comte's importance and influence are hollow, but also because his views on certain questions are both original and perceptive.

THE AIMS AND NATURE OF SCIENCE: PREDICTION AND EXPLANATION

Throughout his *Cours de philosophie positive* (6 vols., 1830–42) and his *Discours sur l'éspirit positif* (1844), Comte stresses that the central aim of science is prevision or prediction; his constant theme, "tout science a pour but la prevoyance" ([4], vol. I, 1830, p. 63), has often been approvingly quoted by those who see science in purely utilitarian terms. However, it is important to stress that Comte's emphasis on the predictive character of scientific knowledge is not based, as Bacon's similar insistence probably was, on a conception of the scientist as *homo faber.*[6] Comte values theories and laws for their own sake, not because they make possible a world of better gadgets. The control science gives us over nature is important for him, of course, but it is by no means the chief virtue of scientific or 'positive' knowledge.

In fact, Comte has a much more interesting methodological reason for focussing on predictive power, for he sees it as something like a *demarcation criterion* which makes it possible to distinguish 'scientific' domains from 'nonscientific' ones. He uses it, in the first instance, to draw a sharp contrast between scientific theories on the one hand and, on the other, metaphysical or theological systems, prediction providing the "unfailing test which distinguishes real *science* from vain *erudition*" ([7], p. 16). But his demarcation criterion is a double-edged device, which Comte also uses to distinguish between legitimate, *systematic* science and the Baconian-inspired compilation of disconnected facts which are just as sterile predictively as metaphysics: "All *science* consists in the coordination of facts; if the diverse observations were entirely isolated, there would be no science."[7] To treat a mere catalogue of facts as a science is "to mistake a foundation for an edifice" ([4], vol. III, 1838, p. 11). Hence it is crucial to avoid thinking that Comte's "principle of prevision" was designed to distinguish between meaningless and meaningful statements, for there are many statements of fact (e.g., "This page is white" or "This sample is sulphur") which, though meaningful, are not scientific (i.e., predictive) in the full sense of the word. The problem of distinguishing between meaningful and meaningless statements is, on Comte's view, a genuine one, which he does try to resolve (as we shall see later) but he sees that as a different problem from the one of distinguishing scientific from nonscientific knowledge.

It is important to realize why it was crucial for Comte to find some differentiating or defining characteristic of scientific knowledge. From the time of Plato, science had usually been distinguished from nonscience primarily in terms of its certainty and infallibility.[8] Even among those who felt that scientific principles could not be absolutely established, it was usually maintained that scientific knowledge consisted only of those statements which were discovered by induction, which were highly probable, and which had 'moral' certainty. In sharp contrast to many of his predecessors, Comte offers an explication of 'scientific' which makes no reference to the truth – or probability – status of theories, and which does not require that scientific theories be generated by induction. He wants to say that a statement is scientific so long as it makes *general* claims about how nature behaves, claims which are capable of being put to experimental test. It is thus *testability* and *generality* which distinguish scientific propositions from nonscientific ones, not how they are discovered nor how certain they are.

Comte's abandonment of traditional demarcation criteria is a significant and, very probably, an influential departure. With Comte's new criterion, it becomes possible to determine whether a statement is scientific *prior* to any investigation of the evidence which might support or refute it. In the decades after Comte, an increasing number of methodologists similarly argue that a proposition is to be regarded as scientific, not if it is highly probable in the light of known evidence, but rather if it is such that we can imagine ways to put it to empirical test.

Despite its importance as a demarcation criterion, predictive ability is not for Comte either the only, or even the chief, characteristic of scientific knowledge. On his view, an equally important function of science is the *explanation* of phenomena. Contrary to what some of his commentators have hinted, Comte was quite prepared to admit that science gives explanations,[9] although he insisted that these are not *ultimate* or *causal* explanations.[10] Indeed, for Comte, explanations and predictions are identical in logical structure, so that what counts as a predictive argument before the fact could equally well count as an explanation after the fact. Every lawlike connection "discovered between any two phenomena enables us both to explain them and to foresee them, each by means of the other" ([7], p. 20).

As for the logical structure of scientific explanation, Comte conceives it as the derivation of an observation statement from other observation statements and certain relevant laws. As he says early in the *Cours de philosophie positive*: "The explanation of facts, reduced to its real terms, will henceforth be only the connection established between the various particular phenomena and some general facts . . . ."[11]

In contrast to his predecessors, who generally insisted that the 'predictions' were inevitably about future events, Comte argued that predictions were not essentially temporal inferences but were distinguished from explanations by an *epistemic* difference. He believed that what was unique to a prediction (that is, what immediately distinguished it from an explanation) was not the 'leap' from the past or present to future but rather, the leap from known to unknown. Comte criticized the usual definition of prediction because it neglected the fact that we often retrospectively 'predict' the occurrence of events which have already happened, such as an eclipse in antiquity.[12]

The principal difference between explanation and prediction, therefore, is this: In an explanation, the initial conditions and the result (the explanandum) are known and it is only a matter of providing the correct law which links the two. In making a prediction, however, we know certain conditions and certain laws, but are uncertain about some event connected to them. When we already know the explanandum to be true, we have an explanation; when we do not, we are making a prediction. A prediction is thus an *anticipation*, which may turn out to be false; an explanation, on the other hand, involves no direct element of test and is simply a demonstration that what happened could have been expected to happen.[13]

MEANING AND DEMARCATION

Comte stresses prediction so emphatically, not because he thinks it more respectable than, or superior to, explanation, but simply because predictions have the important feature of being corrigible or testable. Predictions permit us to test and to verify our laws in a way that explanations do not. This preoccupation with "*vérification*" is a central theme of Comte's work. On his view, every *meaningful* statement must be "open to a positive verification", for *it is only verifiable statements that are meaningful*. He regards it as a 'fundamental rule' that "any proposition which is not strictly reducible to the simple enunciation of the fact – either particular or general – can have no real or intelligible meaning for us".[14] Strictly speaking, this way of putting the point is not altogether satisfactory, for there are presumably many *false* statements which, though palpably not "enunciations of facts," are nonetheless meaningful.[15] He comes closer to expressing the idea when he points out that nonverifiable or meaningless statements are those which we could never have any legitimate (empirical) grounds for affirming or denying.[16] In discussing Comte's theory of verification, one must be careful to emphasize that this is not verification in the strong (i.e., exhaustive) sense in which the logical positivists were later to conceive it. He does not suggest that a statement, in order to be verifiable, must be such that one can exhaustively test all of its implications.[17] His sense of verification is considerably weaker and vaguer than this.[18] What he really seems to be saying is that meaningful statements must bear on the physical world, i.e., they must be nonanalytic and open to checking, either by refutation or by partial confirmation. There is never a hint that the verification of scientific laws involves checking them in every one of their instantiations.[19] Indeed, he seems to have in mind a rather curious theory of domains. He thinks it is scientific and therefore meaningful to say that the law of gravity applies to the solar system because (although we have not applied it to *every* body in the solar system) we have seen that it applies to various types of bodies in the vicinity of the sun. But, because we do (or, rather, did) not have any way of acquiring evidence about the strength of the gravitational field in the region of (say) the pole star, it would be meaningless (and therefore unscientific) to assert (or to deny) that the motion of that star is subject to the law of gravity, prior to the development of techniques for verifying that it applied in such regions ([4], vol. II, 1835, pp. 256 ff). Comte never provided any clear rule for deciding what would distinguish one 'domain' from another, but he seems to feel that we have an intuition as to what constitutes a change of domain of application.

Like the later positivists, Comte used the requirement of verifiability as a stick with which to beat the metaphysicians. This was particularly important for Comte in the light of his three stages of intellectual development. All of the sciences have themselves gone through periods of domination by 'l'espirit métaphysique,' and Comte persistently invokes his verifiability condition as a pruning device for freeing science of those residual metaphysical elements which have not already been purged. But, while Comte's criterion of meaning is sufficient to reject metaphysics, it is not so rigid (as the logical positivists' criterion originally was) as to render science meaningless. Thus, in a schematic way, we could represent Comte's view of the relationship between the problem of meaning and the problem of demarcation in the following manner:

	Predictive	Nonpredictive
Verifiable	Science	Isolated facts
Nonverifiable	–	Metaphysics

INDUCTION, OBSERVATION, AND LAWS

Like most of his contemporaries, Comte claims that the inductive method *is* the scientific method,[20] although (again like most of them) he never spells out in detail exactly what he means by 'induction'. On the other hand, there are a number of clues sprinkled throughout his works on this subject. Perhaps the most significant is his observation that Mill's inductive logic, as formulated in the latter's *System of logic*, best describes his own (i.e., Comte's) point of view ([7], p. 17). But if he is an inductivist, it is a qualified inductivism at best, and certainly one far removed from the Baconian variety. Comte's most significant departure from traditional approaches to induction was his refusal to require that acceptable theories must be 'generated' by some inductive logic of discovery. Where more orthodox inductivists (e.g., Bacon, Newton, and Reid) had insisted that scientific theories must be inductively arrived at, Comte argues that the *origin* of a theory is irrelevant and that what counts is its confirmation: "whether the process whereby [laws] are discovered is ratiocinative or experimental, their scientific value depends entirely on their conformity . . . with the phenomena" ([7], p. 13). He goes on to say that if "inductive rules" are followed too rigidly, science would be led "to mischievious results by encouraging a purely empirical spirit" ([6], vol. I, p. 419). Moreover, and again in contrast to traditional inductivists, Comte insists that the scientist is not and cannot be a passive observer of nature who gradually absorbs and eventually generalizes the facts that he has assimilated. Comte stresses that scientific observation itself presupposes theory, and suggests that the observer without a preconceived theory is of no significance whatever to science:

> If, on the one hand, every positive theory must be based on observations, it is equally clear, on the other hand, that in order to make an observation, our mind requires some theory. If, in contemplating the phenomena, we did not immediately connect them to some principles, it would not only be impossible to combine these isolated observations and consequently impossible to draw anything useful from them, but we would even be altogether incapable of retaining them; and for the most part we could not even perceive them.[21]

Theories are thus essential not only for connecting (or, as Whewell would say, "colligating") our observations, and deducing consequences from them. They are necessary as well for remembering and even for peceiving the 'facts'.

While stressing that science has an observational foundation, Comte sides with the Kantians and against the traditional empiricists and inductivists who had claimed that observation and theory could be clearly differentiated: "Between Observation and Reasoning there is no absolute separation" ([6], vol. I, p. 404). Comte's repudiation of the traditional empiricist theory of observation and concept formation was quite explicit and overt. Time and again, he says that he is trying to avoid both the narrow limitations of the empiricists and the noncritical spirit of the rationalists, to whom he usually refers as mystics: "It is important therefore to understand quite clearly that the true positive spirit is at bottom quite as far removed from empiricism as from mysticism" ([7], p. 16; see also [6], vol. I, p. 419). Comte's realization that "the human mind could never combine or even collect [*recueillir*] observations unless it were directed by some previously adopted speculative doctrine" ([7], p. 6), is closely bound up with his historicist views on the three stages of human thought. When men first began to speculate on nature, they could have made no progress at all if they had simply relied on observations. But, by observing the world first with the aid of theological and then by means of metaphysical theories ([4], vol. I, 1830, p. 9), they were able to build up a body of observational science which could then be purged of these pre-positive elements. The fact that Comte believed one could separate the valid observational core from the extraneous theoretical trappings that made the observation possible in the first place seems to suggest that although Comte concedes that the act of observing is theory-laden, he still thinks it possible to obtain what amount to theoretically neutral observation statements.[22]

Thus far, I have been using the term 'observation' in its most general sense; but more often – particularily when applied to the logic of science – it has a much more specific sense for Comte. Indeed, it is one of only three possible ways of studying nature, the other two being experimentation and what he calls "comparison":

> The art of observation is composed, in general, of three different operations: (1) observation strictly speaking, that is the direct examination of the phenomenon naturally presented to us, (2) experiment, or the consideration of the phenomenon more or less modified by artificial circumstances which we expressly create in order to explore it more perfectly; (3) comparison, or the gradual consideration of a series of analogous cases in which the phenomenon is further and further simplified.[23]

Certain of these methods will be more appropriate to some sciences than others. Thus, the astronomer can (on Comte's account) only observe; the physicist can both observe and experiment but cannot compare; whereas the biologist and social physicist (or sociologist) rely on comparison as well as observation and experimentation.[24]

But, for all his stress on the various methods of observation, Comte is convinced that science is designed ultimately to dispense with observations almost entirely. Indeed, once we know the laws which govern the universe, and a few appropriate initial conditions, science can move from the laboratory to the armchair, from the tedious method of observation and fact-collecting to the more rapid methods of calculation and ratiocination:

> Facts themselves, properly so called, however exact and numerous they may be, can only furnish the indispensable materials of science . . . true science, far from consisting of bare observations, always tends to dispense as much as possible with them and to substitute for them that rational prevision which is, in all respects, the principal characteristic of the Positive spirit . . . .[25]

The discovery of laws and theories, then, is the *raison d'être* of science: once we know what they are, we can simultaneously generate predictions and dispense with all but a few observations; hence, "science really consists in the laws of phenomena" ([7], p. 16).

Having determined the dominant role of laws in Comtean methodology, it is natural to ask what sorts of laws Comte had in mind. Basically, Comte's answer to this question is the same as Hume's. All laws express either regularities of co-existence or regularities of succession ([4], p. 31). What coexist or succeed are the phenomena. But laws express not mere local regularities, but rather, *invariable* and *universal* regularities between phenomena; "The true positive spirit consists above all in seeing for the sake of foreseeing; in studying what *is*, in order to infer what *will be*, in accordance with the general dogma that natural laws are invariable" ([4], p. 26).

The "general dogma" Comte mentions here, "which is the foundation of the whole Positive philosophy" ([7], p. 17) is rather like Mill's principle of the uniformity of nature. It is, at once, both a presupposition of scientific research and a justification for our generalizing from observed regularities to as yet unobserved cases. But how do we know that all natural laws are invariable? Kant, of course, tried to show that this assumption is not derived from experience but is, crudely put, an inevitable result of the way the observer perceives the world. For Kant, we cannot but perceive all events as connected together in regular cause-effect chains. Comte, however, will not accept this solution to the problem of induction:

> Metaphysicians with their meaningless and confused argumentation have tried to represent [the invariability of natural laws] as a sort of innate, or at least primitive idea; whereas it certainly was only reached by a slow, gradual induction . . . ([7], p. 17).

There is nothing in our minds that would suggest, prior to the acquisition of empirical knowledge about natural phenomena, that physical relations are invariable. It is only when we observe frequent connections of certain events that it occurs to us that all events might be subject to invariable laws. Moreover, whatever the origin of the idea of regularity (whether innate or acquired), it draws its validity not from any *a priori* investigation of the nature of the mind, but rather by being put to repeated tests with phenomena:

> The principle of the invariability of natural laws did not really begin to acquire any philosophical weight until the first investigations of a truly scientific kind succeeded in showing its essential accuracy in respect of a whole class of great phenomena . . . ([7], p. 18).

The suggestion seems to be that the principle of the invariability of natural laws, which is now the license for all our predictions, is warranted by the fact that whenever we have tested laws in the past they have proved to be invariable. To the extent that Comte is concerned with justifying induction – and one must say that this did not seem to him to be an acute problem – his justification is itself an inductive one. All he really wants to say is (a) that the possibility of a successful science presupposes the uniformity of nature or the invariability of natural laws (they amount to the same thing for Comte) and (b) that this presupposition cannot be justified *a priori* but only by empirical tests. It is worth noting that, in a sense, Comte evades the usual circularity associated with so many 'empiricist' discussions of induction. Science begins, he suggests, without assuming the general invariability of laws, but with a much more restricted inductive assumption that *certain* phenomena are invariably related. As we observe more and more phenomena which are subject to such regularities, we conclude that all phenomena obey invariable laws. Thus, *the principle of the invariability of natural laws* – far from being a presupposition of all inductive inferences – *arises as a natural inductive generalization from the success of a number of laws obtained inductively*. Comte never suggests that the principle of invariability can be demonstrated or proven;[26] he is content to point out that there is some inductive evidence for its truth.

On Comte's view, therefore, induction in both logically and epistemologically prior to the principle of the invariability of laws; thus, the principle of invariability could not conceivably be used to establish the validity of inductive inferences since that principle itself depends on earlier inductive inferences. But while invariability does not establish that inductive inferences are truth-preserving, it still has a role to play in increasing our confidence in particular inductive predictions.

It is not coincidential that Comte's views here closely resemble Mill's remarks "On the ground of induction" in his *System of Logic*. While scholars have generally stressed Comte's influence on Mill,[27] it seems to me that on the question of induction and the invariability of laws, there are clear signs of influence in the opposite direction. The first edition of Mill's *System* appeared in 1843. Comte's *Cours*, which predates the *System of Logic* by several years, contains no reference to the rationale of the principle of the invariability of laws. However, in his *Discours sur l'éspirit positif*, published in 1844, Comte discusses this question at some length, immediately after a very laudatory reference to Mill's *System*. It is also significant that Comte and Mill had been corresponding extensively since 1841. (Although judging from the extant correspondence [23], this issue was not directly raised between them.) Under the circumstances, it seems plausible to suggest that

Mill's work drew Comte's attention to the problem of induction, as well as to a possible solution for it, and that Comte freely incorporated Mill's views (as he understood them) into his *Discours*. To my knowledge, the only other place where Comte discusses this problem is in the fourth volume of his *System of Positive Polity* (1854), where he again adopts the Millean view that the principle of the invariability of law "will never admit of deductive demonstration, in as much as by its nature it is itself the common basis of all Positive inductions. It will always rest on convictions of an essentially inductive character. . . ."[28]

Although science is thus inductive at its foundations, Comte appreciates the considerable role that deduction plays in it.[29] Indeed, given Comte's views on the dominant role played by prediction, it is natural that he would maintain that deduction was a vital part of scientific method; as Comte himself observed, we only make inductions in order to be in a position to make useful deductions; "Induction is always employed with the view of ultimately deducing" ([6], vol. I, p. 431). Comte thus hoped to steer a midle course "between the two opposite dangers of Mysticism and Empiricism, to which all investigations are liable until the deductive and inductive processes have been properly adjusted" ([6], vol. I, p. 419). Even granting the strong empiricist bias in his work, it should be clear that Comte was no naive inductivist. Recognizing that theory influenced and guided observation and that extensive use could be made of deductive techniques in natural science, he preferred theories to facts and readily conceded that laws go far beyond any direct evidence we can have for them.

HYPOTHESES AND THE THEORY OF 'LOGICAL ARTIFICES'

In an important chapter of the second volume of the *Cours*, Comte devotes a long section to the *Théorie fondamentale des hypothèses* ([4], vol. II, 1835, pp. 433–54). It is this section, more than any other, which undermines the traditional reading of Comte as a narrow-minded empiricist, altogether opposed to hypotheses. Comte begins by suggesting that there are only two means of determining the "real law" of any phenomenon: either by studying the phenomenon inductively and discovering its law, or by deducing the law from a more general one which is already known. But Comte points out that neither method would be possible if we did not first make tentative hypotheses about the form the law will take. We must begin, he insists,

by anticipating the results by making a provisional supposition – at first essentially conjectural – even with regard to some of the very notions which constitute the final object of our research. Hence, the strictly indispensable introduction of hypotheses in natural philosophy.[30]

If we followed Newton's notorious injunction "hypotheses non fingere", the "effective discovery of natural laws would be impossible in any complicated cases, and tedious in every case".[31] Indeed, all the laws of nature "are never anything but hypotheses confirmed by observation".[32] In an obvious way, therefore, hypotheses are for Comte the most fundamental ingredient in the scientific investigation of nature; they are not only essential in the formation of general conclusions, but are also a prerequisite to the execution of any significant experiment.

Accustomed as we are to taking for granted that hypotheses are essential to scientific inquiry, it is easy to underestimate the *historical* importance of Comte's espousal of hypotheses. In the context of early 19th-century thought, Comte's repeated stress on the indispensible role of conjecture is of some considerable significance. Throughout the 18th century, largely as a result of an all-too-literal reading of Newton and Bacon, most scientists and philosophers of science thought it possible to build up scientiic laws and theories inductively or analogically, without using (as they were commonly called) "preconceived hypotheses".[33]

Even among Comte's contemporaries, there were several methodologists (e.g., Ampère, Baden Powell, and J. S. Mill, to name only a few) who believed scientific theories could be constructed without recourse to the method of hypothesis. Thus, Comte's insistence on the necessity of conjecture in scientific method was an important and influential insight.

But in insisting on "the strictly indispensable introduction of hypotheses in natural philosophy," what sort of hypotheses is Comte requiring (and allowing)? A partial answer is provided by what he calls "the fundamental condition for any acceptable hypothesis". Because this condition has given rise to a great deal of confusion, I think it is worth quoting in its entirety:

This condition, hitherto vaguely analysed, consists in conceiving only those hypotheses which are susceptible, by their very nature, of a positive verification, sooner or later, but always inevitably, and whose degree of precision accords exactly with what the study of the corresponding phenomena allows of. In other words, truly philosophical [viz., scientific] hypotheses ought always to have the character of simple anticipations of that which experiment and reason could show immediately, if the circumstances of the problem had been more favorable.[34]

There are several distinct points being made here. Comte begins by requiring that hypotheses must be *verifiable* (in the sense discussed above). As we have seen, this much is required if hypotheses are even to be considered as meaningful. He then specifies two less straightforward conditions: (1) that the precision of the hypothesis be no greater than an empirical analysis of the phenomena in question could warrant, and (2) that the hypothesis must be such that it is no more than an *anticipation* of what 'experiment and reason' could establish directly 'if the circumstances of the problem had been more favorable'. In a sense, the first condition could be said to follow from the requirement that an hypothesis must be verifiable. If we state an hypothesis with a limit of error smaller than existing instruments can measure, then there is an obvious sense in which that precise hypothesis cannot be verified to the exclusion of other equally precise hypotheses similar to it. What we are verifying instead is a much less precise hypothesis. But what are we to make of the second condition? The usual interpretation goes something like this: We are warranted in making hypotheses and conjectures, but only so long as they deal only with 'observable entities', i.e., only so long as they could become phenomenal laws as a result of further investigation. On this account, it is not permissible to make any hypothesis which has (extra-logical) terms in it which refer to unobservable entities. Every hypothesis, in short, must be such that if confirmed it would be an "observational" law of physics.[35]

If this conservative, correlational interpretation (which I shall refer to as CI) is correct,[36] Comte's concessions to the hypothetical method are rather half-hearted. Although he permits us to make conjectures about the correlations between phenomena, we can never make hypotheses about unobservable entities. On this reading, the hypotheses of Dalton on the atom, of Young and Fresnel on the nature of light, of Carnot on heat, of Faraday and Maxwell on the electromagnetic aether would all be excluded as unscientific. More generally, any conjectures that postulated either hypothetical constructs or intervening variables would be declared unscientific. Comte would have to be seen as a philosopher, who allowed for inter-phenomenal hypotheses,[37] but who rejected any hypotheses dealing with imperceptible particles, fluids or forces. Although there is not much direct support for CI, there is some indirect evidence that supports this reading of Comte, at least superficially. (It is well known, for instance, that Comte rejected *both* the corpuscular and the undulatory theories of light. It is also true that he urged his readers not to seek for the underlying causes of phenomena.)

Notwithstanding all this, I believe there are grounds for thinking that Comte's attitude to hypotheses was rather more liberal than CI suggests. We must recall that he is speaking in the quotation cited above about the verifiability of hypotheses. As I suggested earlier, a verifiable statement for

Comte, is one to which experience could make a difference, either by confirming it or refuting it. Clearly, phenomenal hypotheses are verifiable in this sense. But so are a number of nonphenomenal hypotheses, such as the atomic theory. That theory entails, if indirectly, a great many things about the behavior of observable entities (e.g., the laws of definite and multiple proportions). It leads to predictions which can be either confirmed or refuted and is thus verifiable in Comte's sense.

Now, we seem to have a problem on our hands. If Comte means us to take his principle of verifiability seriously – as I think he does – then we must admit as meaningful *any* hypothesis that experience could confirm or refute, either directly or indirectly. However, if a hypothesis can only contain observable predicates, then many verifiable hypotheses must nonetheless be excluded. It is precisely at his point that Comte's reference to 'reason' in the passage cited above becomes significant. We could never, by merely generalizing data from experiment, arrive at a theory which dealt with atoms. But, given an atomic theory like Dalton's, it is possible for reason to deduce experimental consequences from it which can be verified; and on this more liberal interpretation of Comte's fundamental requirement (which I shall call LI), the atomic theory would be an acceptable hypothesis.

In support of LI, one can point out that Comte, in methodological as well as scientific contexts, accepts many hypotheses which are not purely phenomenal ones, including the atomic theory. In his *System of Positive Polity*, for instance, he explicitly endorses the atomic theory and, in so doing, offers the following remark on hypotheses "It is consistent with sound reasoning to make use of any hypotheses that will assist thought, provided always that they be not inconsistent with what we know of the phenomena" ([6], vol. I, p. 421). Now, if CI is sound, this new stipulation is obviously inconsistent with the 'fundamental condition' cited above. We would therefore have to assume that Comte's ideas on the nature of hypotheses underwent radical modifications between the *Cours* and the *System*, which is, of course, possible. However, if LI is correct, then the theory of hypothesis which Comte espouses in the *System* is fully compatible with his discussion of hypotheses in the *Cours*.

The chief evidence usually cited for CI is Comte's denunciation of corpuscular and wave theories of light. But, a closer examination of his views on those theories reveals that he does not reject them because they postulate unobservable entities (i.e., not because they go beyond correlations of facts), but for quite different reasons. For one thing, Comte thinks that there is some experimental evidence inconsistent with each hypothesis; moreover, by and large, he says that their empirical import is the same in the sense that they both entail all the known laws of optics.[38] As they are false in certain domains and overlapping in all the others, he felt that little hung on the controversy between the partisans of the two opinions. If he had realized that the two theories were *not* observationally equivalent, he presumably would have been more receptive to taking them seriously.[39]

Even if it is plausible to liberalize the usual account of Comte's theory of hypothesis, there are still a great many possible hypotheses which it excludes.[40] Specifically, any hypothesis which deals with first causes or final causes is certainly excluded by his fundamental condition. All he really wants to exclude among the set of possible hypotheses are those which violate his condition of positive knowledge: "If one presumes to reach by hypothesis what is in itself radically inaccessible to observation and reasoning, the fundamental condition will go unheeded and the hypothesis, going beyond the true scientific domain, necessarily becomes injurious."[41] But, beyond that, he is quite prepared to endorse what he calls "l'usage rationnel des hypothèses" ([4], vol. II, 1835, p. 465) in all domains of scientific inquiry.

When we have two or more hypotheses equally compatible with the phenomena, Comte suggests that we should prefer the simplest one. Indeed, "Law One" of Comte's first philosophy expresses "the rule that we should in all cases form the simplest hypothesis consistent with the whole of the facts to be presented."[42] Comte's insistence (to be found at several places in his works) on simple hypotheses which are *consistent* with the evidence seems, at first glance, to undermine views he has expressed elsewhere. After all, the hypotheses of demons, angels, entelechies, and the like, are presumably consistent with empirical evidence; they must be, for on Comte's view such hypotheses are compatible with any empirical state of affairs whatever. It should be remembered, however, that Comte's stress on simplicity is meant to be taken in conjunction with, and not as an alternative to, his fundamental condition that hypotheses must be verifiable. Combining the two, we might say that Comte's injunction is *always to form the simplest, verifiable hypothesis compatible with all the known evidence*. Thus, far from excluding hypotheses dealing with entities that are not directly observable, he expressly allows for them, provided they satisfy certain plausible, if rather vague, conditions.

There is an important qualification which should be added here. Comte perhaps is, as I have suggested, as liberal as one might like when it comes to the kinds of hypotheses which he will allow the 'positive' scientist to entertain. However, when it comes to the *interpretation* of those hypotheses, he adopts a rather different position than certain scientific "realists" and "reductionists" would like. Basically, he is prepared to consider hypotheses apparently about atoms, invisible fluids, etc. insofar as they aid, and are necessary for, the correlation of *observation statements*; but he insists that they be treated as fictions or (to use anachronistic language) as intervening variables. In other words, the scientist who uses an atomic or molecular theory is not really assuming the existence of atoms or molecules (unless, of course, he has some way of detecting them), but only adopting a language which permits the unification of many laws that, in the absence of such theories, would appear disconnected and fragmentary. Thus, Comte suggests that we should treat the atomic hypothesis as a "logical artifice" and not as a physical discovery; we must restrain "our tendency to endow all subjective creations with objective existence, as though they represented some external reality" ([6], vol. I, p. 421). He explains that his original 'tough line' on hypotheses was adopted for fear that, if scientists were allowed to make hypotheses about unobservable entities, they would be deluded into thinking that the success of those theories was a sign that their constituent entities had physical reality. As he was well aware, the psychological temptation to attribute reality to useful fictions was painfully difficult to resist:

> Is it even possible, after having adopted a notion which admits of no verification, to use it continually, letting it freely mingle with real ideas, without involuntarily attributing to it an effective existence . . . . ? ([6], vol. II, p. 442).[43]

It is natural to conclude from passages such as these that Comte was convinced that the meaning of an hypothesis was to be found by *reducing* – the term is his – that hypothesis to a statement expressible in the language of observation. However, this is not to say that Comte proposed replacing hypotheses (à la Ramsey or Craig) by the statements which were their observational translations. Comte recognized that many hypotheses achieved a kind of conceptual economy which their observational analogues lacked. Consider, again, the atomic theory. If one wants to determine its existential commitments, then one may reduce it to its observational implications. But, the atomic theory groups together a number of observational laws and exhibits their interrelatedness. If we were to abandon the atomic hypothesis altogether, this could only be, as Comte was well aware, at the expense of theoretical unity.

But another major problem still remains: granting that some hypotheses are designed solely to correlate our laws and do not assert the objective existence of anything unobservable, are there not some hypotheses which probably represent things as they really are, even though they cannot now be directly verified?[44] Let us return to the example of the atomic theory. We must grant immediately that the atomic hypothesis was *generally* used by Comte's contemporaries as a useful fiction.[45] But there were atomists who, while conceding that atoms could not then be directly observed, nonetheless maintained that there really were atoms and expressed the conviction that sooner or later more or less direct evidence could be found that would demonstrate their existence. It is very difficult to guess what Comte would have made of this doctrine, for he does not seem ever to consider the possibility that a scientist might *legitimately* postulate the existence of an unobservable entity. However, because he consistently treats hypotheses such as the atomic theory as fictions rather than as putatively true statements, one suspects that he may have envisaged a rule roughly like the following: If the terms in an hypothesis refer to directly observable entities, then the hypothesis is either true or false and is committed to the existence of the entities it describes; if the terms in the hypothesis refer to entities not directly observable, then it neither is true or false, only more or less useful, and its constituent terms cannot be regarded as referential. Since the distinction between directly observable and indirectly observable corresponds closely to the more recent distinction between directly and indirectly verifiable (as in Ayer, for instance), one can restate the rule: if an hypothesis is only indirectly verifiable then it is a fiction, a 'logical artifice', but nonetheless scientific. Comte evidently did not consider the important question of whether indirectly verifiable hypotheses might become in time directly verifiable ones. It is probably just as well from his point of view that he ignored this question, for one suspects he could have found himself in the awkward position of saying that the referential status of an hypothesis changes radically simply as a result of technological and instrumental breakthroughs.

Inevitably, given the present state of Comtean scholarship, this investigation has been highly tentative and sketchy. Clearly, too, it poses exegetical problems rather than offering a definitive resolution of them. Moreover, the investigation has been ahistorical in the sense that I have not been able to explore here the *evolution* of Comte's thought on these questions from 1830 to about 1850, nor have I been able to consider what I suspect is probably a profound debt on Comte's part to his predecessors in these matters. Although this preliminary survey has compressed the arguments of a dozen volumes into a few pages, it should at least make clear that Comte's theory of scientific method deserves to be analyzed rather more carefully than historians of philosophy of science have recognized.

NOTES

1 To get a sense of the magnitude of Mill's debt to Comte, it is only necessary to look at the first edition of [22]. In later editions of that work, Mill systematically deletes many allusions to Comte. Thus, in 1843, he asserts that *vis-à-vis* the place of hypothesis in science "M. Comte . . . of all philosophers seems to me to have approached the nearest to a sound view of this important subject" ([22], vol. II, 1843, p. 17). In later editions, although Mill continues to advocate the Comtean theory of hypothesis, he deletes this laudatory reference entirely. (There are similar passages collected together as the first appendix to [30].)

2 Cf. [17], [18], and, for another 'positivist,' [2].

3 Cf. Whewell's two major essays on Comte: ([31], vol. II, pp. 320–33, and [32]).

4 However, if my assessment of Comte has any truth in it, the members of the Vienna Circle knew Comte only second-hand and in a very diluted form [15].

5 The only lengthy exegetical treatments of Comte's philosphy of science which I have found are [8], [9], and [16]. Despite their many merits, none of these works contains a critical discussion of Comte's account of such methodological concepts as law, verification or hypothesis.

6 Hayek, in his otherwise excellent study of Comte [11], lays far too much emphasis on Comte's alleged commitment to practical, useful knowledge. (See especially pp. 182 ff.)

7 "Tout science consiste dans la coordination des faits; si les diverses observations étaient entierèment isolées, il n'y aurait pas de science" ([4], vol. I, 1830, p. 131).

8 This innovation is very closely linked to the declining fortunes of the 'logic of discovery'; see Chapter 11 below.

9 Meyerson, for instance, claims that "Comte's attitude in regard to explanatory theories necessitates withdrawing them entirely from science" ([21], p. 45).

10 As Mill pointed out, it was not to the term 'explanation' that Comte objected, but to the term 'cause', understood in the sense of an efficient cause ([10], p. 223). Unlike Mill, Comte was usually unwilling to use 'cause' to designate an invariable antecedent, and believed that the 'inquiry after *causes*' ought to be abandoned altogether ([7], p. 13).

11 "L'explication des faits, réduite alors à ses termes réels, n'est plus desormais que la liaison établie entre les divers phénomènes particuliers et quelques faits généraux, dont les progres de la science tendent de plus en plus à diminuer le nombre" ([4], vol. I, 1830, p. 5).

In [6], Comte similarly observes that "explanation of phenomena is nothing but the process of connecting them together by [laws of] similitude or succession . . . " ([6], vol. I, p. 409).

12 " . . . prevision, in the scientific sense of the word, is not confined to the future; it may evidently be used also of the present and even of the past" ([7], p. 21). Although Comte probably was, as I suggested, the first to make this point specifically about predictions, it had been anticipated some years earlier by Bailey who argued that inductive inferences (he does not use the term 'predictions') refer not to arguments from past to future but from known to unknown. (See [1], especially Essay III, Chapter 2.) Mill made a similar point about induction in [10], Chapter 3 of Book 3, and this may well have been Comte's source, particularly in the light of evidence I present below about Mill's influence on Comte.

[13] Although Comte is usually careful, when explicating the concept of explanation, to treat it simply as an inference from laws and observations to further observations, he occasionally lapses into more traditional modes of speech. Thus, in volume two of the *Cours*, at one point he says that a "true *explanation*" is provided "par une exacte comparaison des moins connus aux plus connus" ([4], vol. II, 1835, p. 247). (Generally, however, he scrupulously avoids the language of Aristotle's *Posterior Analytics* when dicussing methodological matters.)

[14] "Toute proposition qui n'est strictement réductible à la simple énonciation d'un fait, ou particulier ou général, ne peut nous offrir aucun sens réel et intelligible" ([7], pp. 12–13). By "general facts" Comte presumably means lawlike statements.

[15] It is likely that, when speaking of propositions being reducible to a fact, Comte means something like 'reducible to a statement about matters of fact', which still permits false propositions to be meaningful.

[16] " . . . any conceptions formed by our imagination, if of such a nature as to be necessarily inaccessible to observation, are no more capable of direct negation than of direct affirmation" ([7], p. 43).

[17] In [27], Popper mistakenly asserts that Comte (by allegedly requiring *exhaustive* verifiability) "overlooked the problem of universality or generality" ([27], vol. II, p. 298). As Popper himself concedes, however, he can assimilate Comte's theory of meaning to that of the logical positivists *only by omitting certain crucial passages* in Comte's discussion of meaning.

[18] Peirce, who credits Comte with having been the first to require that scientific hypotheses must be verifiable, is justifiably dismayed at the ambiguity of Comte's notion of verification. On Peirce's view, Comte's requirement of verifiability was not accepted by his contemporaries "chiefly because Comte did not make it clear, nor did he apparently understand, what verification consisted in" ([25], 7.91). Nonetheless, Peirce still believed "that such theories as that of Comte and Poincaré about verifiable hypotheses frequently deserve the most serious consideration". I believe that Peirce's attribution of priority to Comte is probably mistaken.

[19] The use of the term 'vérification' to denote something like confirmation rather than *exhaustive* verification was not uncommon in the 19th century. The French logician and philosopher of science, Duval-Jouve, uses it precisely in the former sense in [10]. (See especially [10], p. 192 ff., where Duval-Jouve argues that all hypotheses must be "verifiable".) There is no definite evidence that Duval-Jouve adopted this use of the term from Comte. Mill, Herschel, and Whewell also use 'verification' as a general term which simply denotes empirical testing. (See, for instance, [22], Book III, Chapter XI, page 3.) None of these other writers who speak of verification go as far as Comte does in identifying verifiability with meaning.

[20] [7], p. 17n. Elsewhere he wrote: "The construction of Inductive Logic, of which antiquity had scarce any conception, is the principal feature of the modern mind" ([6], I, p. 419).

[21] "Car, si d'un côté, toute théorie positive doit nécessairement être fondée sur les observations, il est également sensible, d'un autre côté, que, pour se livrer à l'observation, notre esprit a besoin d'une théorie quelconque. Si en contemplant les phénomènes, nous ne les rattachions point immédiatement à quelques principes, non-seulement il nous serait impossible de combiner ces observations isolées, et par conséquent, d'entirer aucun fruit, mais nous serions même entièrement incapables de les retenir; et, le plus souvent, les faits

resteraient inaperçus sous nos yeux" ([4], vol. I, 1830, pp. 8–9). In [7], he similarly remarks: "while on the other hand it is quite true, as we moderns proclaim, that no solid theory can be built except on a sufficient combination of suitable observations, it is on the other hand incontestable that the human mind could never combine or even collect such observations unless it were directed by some previously adopted speculative doctrine or theory" ([7], p. 6).

22 See especially Comte's discussion of optics in the 28e Leçon of [4], where he claims that such neutral observation statements exist: "En optique, par exemple, le mot rayon, si bien construit pour l'hypothèse de l'émission continue aujourd'hui a être employé par le partisans des ondulations: il ne serait pas plus difficile, de lui attacher un sens indépendant d'aucune hypothèse, et simplement relatif au phénomène" ([4], vol. II, 1835, pp. 444–45). Whewell completely repudiated Comte's claim on this matter and insisted that it was impossible even to express the phenomenal laws of optics without presupposing the undulatory hypothesis (see [16], vol. II, pp. 326–327; and Brewster's reply in [3]).

23 "Notre art d'observer se compose, en général, de trois procédés différens: 1° l'observation proprement dite, c'est-à-dire l'examen direct du phénomène tel qu'il se présente naturellement; 2° l'expérience, c'est-à-dire la contemplation du phénomène plus ou moins modifié par des circonstances artificielles, que nous instituons expressément en vue d'une plus parfaite exploration; 3° la comparaison, c'est-à-dire la considération graduelle d'un suite de cas analogues, dans lesquels le phénomène se simplifie de plus en plus" ([4], vol. II, 1835, p. 19).

24 I shall say nothing here about a fourth method which Comte claims to have invented and which he believes to be crucial to the social sciences – the so-called 'historical method'.

25 "Les faits proprement dits, quelque exacts et nombreux qu'ils puissent être, ne fournissent jamais que d'indispensables matériaux . . . la véritable science, bien loin d'être formée de simples observations, tend toujours à dispenser, autant que possible, en y substituant cette prévision rationelle, qui constitue, à tous égards, le principal caractère de l'éspirit positif . . . " ([7], p. 16). As early as 1830, Comte had come to such a conclusion. In vol. I of the *Cours*, for instance, he says: "On peut même dire généralement que la science est essentiellement destinée à dispenser . . . de toute observation directe, en permetant de déduire du plus petit nombre possible de données immédiates, le plus grand nombre possible de résultats" ([4], vol. I, 1830, p. 131).

26 As he wrote in 1851 to Papot: "the most fundamental dogma of the whole of positive philosophy, that is to say, the subjection of all real phenomena to invariable laws, only results with certainty from an immense induction, without really being deducible from any notion whatever" (quoted in [16], p. 85).

27 For instance, [24] and [30].

28 [6], vol. IV, p. 169. In a discussion of the principle of uniformity or invariability, A. D. Ritchie – with a remarkable combination of facetious history, poor taste, and good fun – writes: "The favourite device for saving the face of Induction is the Principle of the Uniformity of Nature. I am not aware when the Principle was first introduced to an admiring world; but, as it has a pleasantly Victorian flavour about it, I am inclined to conjecture that it made its debut at the Exhibition of 1851, when it was awarded a Silver Medal" ([14], p. 91).

29 Offering an up-dated version of Aristotle, Comte says that induction must be used to

establish the axioms of any science, while deuction established the theorems: "Les avantages respectifs de ces modes [induction et déduction] varient beaucoup le nature de cas scientifiques: il faut, autant que possible, préférer habituellement la déduction pour les recherches spéciales, et réserver induction pour les seules lois fondamentales ... Si l'abus de la seconde tend directement à faire dégénérer la science en une confuse accumulation de lois incohérentes, il est pareillement incontestable, que l'emploi exagéré de la premier altère nécesairement l'utilité, la netteté, et même, la realité de nos spéculations quelconques" ([4], vol. VI, 1842, p. 718).

As he says elsewhere "il ne peut exister que deux moyens généraux ... en un mot, l'induction ou la déduction" ([4], vol. II, 1835, p. 357).

30 " ... si l'on ne commençait souvent par anticiper sur les résultats, en faisant une supposition provisoire, d'abord essentiellement conjecturale, quant à quelques unes des notions mêmes qui constituent l'objet final de la recherche. De là, l'introduction, strictement indispensable, des hypothèses en philosophie naturelle" ([4], vol. II, 1835, p. 434).

31 " ... le découverte effective des lois naturelles serait évidemment impossible, pour peu que le cas présentât de complication; et, toujours, le progres réel serait, au moins, extrêmement ralenti" ([4], vol. II, 1835, p. 434).

32 Quoted in [8], p. 101.

33 For the relevant background, see Chapters 7 and 8 above.

34 "Cette condiion, jusqui'ici vaguement analysée, consiste à jamais imaginer que des hypothèses, susceptibles, par leur nature, d'une vérification positive, plus ou moins éloignée, mais toujours clairement inévitable, et dont le degré de précision soit exactement en harmonie avec celui que comporte l'étude des phénomènes correspondans. En d'autres termes, les hypothèses vraiment philosophiques doivent constamment présenter le caractère de simples anticipations sur ce que l'expérience et le raisonnement auraient pu dévolier immédiatement, si les circonstances du problème eussent été plus favorables" ([4], vol. II, 1835, pp. 434–5).

35 Among those who read Comte as advocating a view similar to this, see [8], pp. 89 ff.

36 This reading of Comte's 'fundamental condition' was first expressed by Whewell and Mill, both of whom criticize Comte for taking too limited a view of the scope of scientific hypotheses. Peirce and Chauncey Wright, along with more recent commentators, have accepted this reading of the fundamental condition. Peirce, for instance, without any discussion of the relevant texts, charges that "Comte's own notion of a verifiable hypothesis was that it must not suppose anything that you are not able directly to observe" ([25], 5.597).

37 By the phrase 'inter-phenomenal hypotheses" I simply mean hypotheses whose (non-logical) predicates are drawn exclusively from observation.

38 See above, note 22.

39 Meyerson suggests yet another alternative. He claims that Comte in endorsing the atomic theory "has yielded less to his principles and more to his powerful scientific instinct" ([21], p. 61). Meyerson thereby ignores the fact that Comte had methodological as well as scientific reasons for accepting the atomic hypothesis.

40 To this extent, I believe Levy-Bruhl is wrong when he claims that: "Far from giving too small a share to hypothesis, like Bacon, Comte would rather recur the reproof of having given it too large a one" ([16], p. 148).

41 " ... si l'on prétendait atteindre par l'hypothèse ce qui, en soi-même, est radicale-

ment inaccessible à l'observation et au raisonnement, la condition fondamentale serait méconnue, et l'hypothèse, sortant aussitôt du vrai domaine scientifique, deviendrait nécessairement nuisible" ([4], vol. II, 1835, p. 435).

42 ([6], vol. IV, p. 154). This "law" was first formulated by Comte in [4], p. 58.

43 "Est-il vraiment possible, après avoir adopté une notion qui ne comporte aucune vérification, d'en faire un usage continuel, de la mêler intimement à toutes les idées réeles, sans être jamais involontairement entrainé à lui attribuer une existence effective . . . " ([4], vol. II, 1835, pp. 441–2).

44 Comte's Scottish contemporary, Macquorn Rankine, called these two forms of hypotheses 'objective' and 'subjective' hypotheses respectively ([28], p. 210).

45 See, for instance, [12], pp. 20 ff.

REFERENCES

[1] Bailey, S., *Essays on the Pursuit of Truth*, London, 1829.

[2] Bernard, C., *Introduction à l'Etude de la Médecine Expérimentale*, Paris, 1865.

[3] Brewster, D., Review of 1840 edition, volumes I and II of [31], *Edinburgh Review*, 1842, 482–91.

[4] Comte, A., *Cours de Philosophie Positive*, 1st ed., 6 vols., Paris, 1830–42.

[5] Comte, A., *Géométrie Analytique*, Paris, 1843.

[6] Comte, A., *System of Positive Polity*, 4 vols. (trans. Bridges *et alia*), London, 1875–7.

[7] Comte, A., *Discours dur l'Espirit Positif*, Paris, 1844.

[8] Delvolvé, J., *Réflexions sur la Pensée Comtienne*, Paris, 1932.

[9] Ducassé, P., *Méthode et Intuition chez Auguste Comte*, Paris, 1932.

[10] Duval-Jouve, J., *Traité de Logique, ou Essai sur la Théorie de la Science*, Paris, 1844.

[11] Hayek, F., *The Counter-Revolution of Science*, London, 1955.

[12] Knight, D., *Atoms and Elements*, London, 1967.

[13] Laudan, L., 'The Nature and Sources of Locke's Views on Hypotheses', *Journal of the History of Ideas* 28 (1967), 211–23.

[14] Laudan, L., 'Thomas Reid and the Newtonian Turn of British Methodological Thought', in *The Methodological Heritage of Newton* (ed. R. Butts), Toronto, 1970.

[15] Laudan, L., *Progress and Its Problems*, Berkeley, 1977.

[16] Lévy-Bruhl, L., *The Philosophy of Auguste Comte* (trans. F. Harrison), New York, 1903.

[17] Littré, E., *De la Philosophie Positive*, Paris, 1845.

[18] Littré, E., *Auguste Comte et la Philosophie Positive*, 2nd ed., Paris, 1864.

[19] Locke, J., *Essay Concerning the Human Understanding*, London, 1690.

[20] Martin, B., *Philosophical Grammar*, Reading, 1748.

[21] Meyerson, E., *Identity and Reality* (trans. Loewenburg), New York, 1962.

[22] Mill, J. S., *A System of Logic, Ratiocinative and Inductive*, 2 vols., London, 1843.

[23] Mill, J. S., *Lettres Inédites de John Stuart Mill à Auguste Comte* (ed. L. Lévy Bruhl), Paris, 1899.

[24] Mueller, I., *John Stuart Mill and French Thought*, Urbana, Ill. 1956.

[25] Peirce, C. S., *Collected Papers* (ed. P. Weiss *et alia*), 7 vols., Cambridge, Mass., 1932–58.
[26] Popper, K., *Conjectures and Refutations*, London, 1963.
[27] Popper, K., *The Open Society and Its Enemies*, 2 vols., Princeton, 1963.
[28] Rankine, J., *Miscellaneous Scientific Papers*, Edinburgh, 1894.
[29] Reid, T., *Works* (ed. Hamilton), Edinburgh, 1858.
[30] Simon, W., *European Positivism in the Nineteenth Century*, Ithaca, 1963.
[31] Whewell, W., *The Philosophy of the Inductive Sciences, Founded Upon Their History* (eds. Buchdahl & Laudan), 2 vols., London, 1968.
[32] Whewell, W., 'Comte and Positivism', *Macmillan's Magazine* 13 (1866), 353–62.

CHAPTER 10

WILLIAM WHEWELL ON THE CONSILIENCE OF INDUCTIONS

Few scholars would deny that William Whewell ranks among the major figures in 19th-century philosophy of science. His *Philosophy of the Inductive Sciences* and his later *Philosophy of Discovery* remain among the classics of scientific methodology. The bulk of the scholarship devoted to Whewell has tended to stress the idealistic, anti-empirical temper of Whewell's philosophy. For instance, the only two monograph-length studies on Whewell, Blanche's *Le Rationalisme de Whewell* (1935) and Marcucci's *L'Idealismo' Scientifico di William Whewell* (1963), are, as their titles suggest, concerned primarily with Whewell's *departures* from empiricism. Other studies, especially those of Robert Butts, have given prominence to neo-Kantian elements in Whewell. Particularly in recounting Whewell's famous dispute with Mill, it has proved tempting to parody Whewell's position in the debate by treating the controversy as a straightforward encounter between an arch-empiricist and an arch-rationalist. There is, however, a danger that an emphasis on the rationalist and *a priori* elements in Whewell's philosophy may well obscure the unmistakable empirical emphasis in Whewell's theory of science. I think it is time to begin to redress the balance, by focussing attention on the significant 'empiricist' strains in Whewell's theory of science. One of the most important of those strains is connected with the operation which Whewell calls 'the consilience of inductions'.

Indeed, of all the fanciful neologisms which Whewell coined (including 'the colligation of facts', 'the explication of conceptions', 'the decomposition of facts' and 'the superinduction of conceptions'), none denoted a more fertile methodological process nor a more important doctrine in Whewell's methodology, than '*the consilience of inductions*'. This fact alone would more than justify a close scrutiny of this doctrine. However, a fuller understanding of the nature of Whewell's views on consilience is not only vital to a comprehension of his philosophy of science, but it also the key to his historiography of science, for it is largely in terms of consiliences that Whewell formulates his theory of the *progressive* nature of scientific growth and evolution. This chapter seeks to give a brief explication of Whewell's notion of the consilience of inductions, along with an assessment of the role which consilience and related concepts played in Whewell's history and philosphy of science.

In the fourteenth aphorism concerning science in the *Philosophy of the Inductive Sciences* (1840), Whewell offers what is probably his briefest characterization of the nature of consilience:

> The Consilience of Inductions takes place when an Induction, obtained from one class of facts, coincides with an Induction, obtained from another class. This Consilience is a test of the truth of the Theory in which it occurs.[1]

To paraphrase, if two chains of 'inductive reasoning' from seemingly different classes of phenomena lead us to the same 'conclusion', a consilience of inductions has occurred. In the light of this aphorism, there are two significant questions to ask initially about the consilience of inductions: (a) precisely what is involved in a consilience?[2] and (b) why should a successful consilience count as "a test of the truth of the theory?"

Because the notion of consilience is so clearly connected with Whewell's doctrine of induction (being effectively the result of two or more inductions leading to the same general proposition), it is important to be clear at the outset about the nature of Whewellian induction. It is generally accepted (as Mill charged at the time) that Whewell radically transformed the traditional meaning(s) of 'induction', i.e., Whewellian induction is neither *induction per enumerationem simplicem* nor is it at all akin to the eliminative induction of Bacon or Mill. On the contrary, induction is seen by Whewell as a *conjectural* process whereby we introduce a new conception, not immediately given 'in' the available evidence, which 'colligates' that evidence, while going beyond it both in generality and degree of abstraction.[3]

Whewellian induction is thus similar to what Peirce was later to call *abduction* or *retroduction,*[4] and consists essentially in finding some general hypothesis which entails the known facts. As Whewell says,

> our Inductive Formula might be something like the following: 'These particulars, and all known particulars of the same kind, are exactly expressed by adopting the Conceptions and Statement of the following Proposition'.[5]

Using Whewell's technical terminology, an induction is generally a successful colligation of facts by a clear and appropriate conception. Adapting Whewell's language to the kind of schema required for a consilience, we might say that *the formula for a consilience of inductions* is as follows:

> These particulars of different types, A_1 & . . . & A_n ($n \geqslant 2$), and all known particulars of the same types, are exactly expressed by adopting the conceptions and statement of the following Proposition: . . .[5]

Given that induction is the formulation of an hypothesis which will explain (or "express") a class of known facts, it follows that a consilience of inductions occurs when we discover that the same hypothesis explains (or expresses) two (or more) classes of facts.[6] Disguised within this terse statement are several important methodological claims. Indeed, by unpacking Whewell's various treatments of consilience, I think we can suggest that a consilience occurs under the following circumstances:

(1) When an hypothesis is capable of explaining two (or more) *known* classes of facts (or laws);
(2) When an hypothesis can successfully *predict* "cases of a *kind different* from those which were contemplated in the formation of our hypothesis";[7]
(3) When an hypothesis can successfully predict or explain the occurrence of phenomena which, on the basis of our background knowledge, we would not have expected to occur.[8]

As these instances suggest, and as I shall argue in detail later, a consilience of inductions occurs when an hypothesis is shown to have achieved a certain minimum of corroborated or confirmed empirical content. More specifically, an hypothesis achieves a consilience when it has shown itself to be capable of successfully explaining either different kinds of facts, or very surprising facts. My interpretation here, and in what follows, is in rather sharp contrast to the view that consilience is *entirely* a matter of increasing empirical content; a consilience of inductions occurring whenever a certain minimal gain in generality has been achieved.[9] On the contrary, I am convinced that Whewell's aim in stressing the consilience of inductions is not to maximize content, but to maximize the confirmation of an hypothesis. Of course, Whewell did believe that in the progressive growth of science, we advance towards theories of greater scope, range and generality. But (and this is crucial) increased generality is only a gain insofar as that greater generality is *experimentally confirmed*. Consilience is, effectively, a criterion of acceptability which stipulates that those hypotheses are most worthy of belief and acceptance which pass empirical tests of the kind sketched above.

To return to the cases outlined there, it is obvious that they are not mutually exclusive; if condition (3) is satisfied, so generally will condition (2).[10] The grounds for distinguishing them as I have done are twofold: (a) that all three constitute slightly different methodological gambits, and (b) that the justification given for each of the three depends upon slightly different considerations.

Take the first case. A consilience of inductions of this type (which I shall hereinafter refer to as CI_1) does not, as CI_2 and CI_3 do, increase the empirical content of our theoretical knowledge. Such consiliences are concerned neither with the predictive ability of our theories nor with their ability to extend our knowledge into new domains. However, what a CI_1 does accomplish is the *formal unification* or *simplification* of our theories and hypotheses. By reducing two classes of phenomena – which had hitherto required separate and (seemingly) independent hypotheses or theories for their explanation – to one general hypothesis or theory, CI_1 clearly achieves a reduction in the theoretical baggage required to 'carry' the known phenomena. The advantages to be gained by this type of consilience are, therefore, formal or systemic, and not empirical, ones.

On the other hand, consiliences of type CI_2 (i.e., (2) above) represent an increase in the explanatory content of our science, insofar as the adoption of the new hypothesis (which exhibits CI_2) permits us to anticipate natural phenomena which were unknown (or at least unconsidered) when the theory was devised. Such fecundity is clearly an important attribute of a new hypothesis and Whewell views it as a strong mark in favor of an hypothesis if it can achieve a CI_2.[11] As Whewell remarks,

> when the hypothesis of itself and without adjustment for that purpose, gives us the rule and reason of a class of facts not contemplated in its construction, we have a criterion of its reality, which has never yet been produced in favour of falsehood.[12]

Like CI_2, consiliences of type CI_3 mark a gain in the explanatory power of our theories. The special strength of the latter type of consilience, however, is that it not only permits us to explain (and predict) more kinds of phenomena than we could previously, it also enables us to test our theory in a uniquely stringent way by seeing whether it successfully predicts phenomena which, in the absence of that theory, we would have regarded as either impossible, inexplicable, or at least highly unlikely. Embodied in CI_3, therefore, is a notion something like what Popper was later to call 'the severity of tests'. If a theory has surprising consequences which are, in fact, corroborated, we are disposed to regard it as a very sound hypothesis.[13] Such, in brief, are the three types of consilience implicit in Whewell's methodology.

Although Whewell talks as if the value of a 'consiliative' hypothesis is that it explains (or expresses and predicts) events of different 'kinds' or from different 'classes of phenomena', the real strength of such an hypothesis is usually that *it shows that events previously thought to be of different kinds are, as a matter of fact, the 'same' kind of event.* For instance, Newton's gravitational hypothesis did not, *in its own terms*, explain 'different' kinds of events; rather, it showed that (for instance) the motion of the moon and the fall of a heavy body on earth were precisely the *same* type of event. It was (in a sense) merely a fortunate accident of historical circumstance that Newton propounded his theory at a time when these were regarded as very different types of phenomena.

Such considerations suggest that there is something essentially *historical* and *relative* about deciding whether a given theory achieves a consilience of inductions. *Vis-à-vis* the Cartesian system, for instance, the Newtonian explanation of planetary and terrestrial motions was *not* consilient (in this respect), for Descartes had regarded both kinds of phenomena as due to the action of roughly similar types of vortices. Thus, the decision as to whether a given theory has achieved a consilience of inductions can only be reached via a careful study of those other theories with which it is competing at a given time. Without a thorough knowledge of historical context, and without overt reference to alternative theories (perhaps merely in the form of so-called background knowledge), it is usually impossible for us to decide whether a given theory achieves a consilience or not. The reason for this is clear. A physical theory itself generally can not express the fact that the phenomena which it "consiliates" are different in kind, for that theory – insofar as it explains them in similar terms – must regard them as phenomena of the same kind. Thus, consilience makes reference to at least two theories. It follows that it is misleading to say absolutely that a theory T has achieved a consilience of inductions; rather, we should say that T, relative to the natural kinds specified in other (competing) theories T_1 & . . . & T_n, has achieved a consilience.

The obvious question to ask is why Whewell regarded consiliences of the type outlined above as of great value. Construed as methodological rules, there is nothing very surprising about CI_1 and CI_2. Both had been frequently cited methodological rules many years before the *Philosophy of the Inductive Sciences.*[14] What is interesting, however, are the reasons Whewell gives for regarding consiliences as extremely important characteristics of scientific hypotheses. As Whewell indicated in the Aphorism quoted above, he considers such consiliences as a "test of the truth of the Theory" in which they occur. Moreover, they are tests in the strong sense of the term; that is, a consilience of inductions constitutes a sufficient and not merely (nor even) a necessary test of the truth of the theory in which it occurs. But why should a consilience, whether it brings about formal unification (CI_1) or an increase in empirical content (CI_2 and CI_3), be regarded as strong indication, let alone a proof that the theory in which occurs is true? It is the answer Whewell gives to such questions which I now want to explore.

In discussing the logic of hypothesis testing in the *Philosophy of the Inductive Sciences* and in the *Novum Organon Renovatum* (1858), Whewell differentiated several *degrees of severity* associated with the tests to which scientific hypotheses could be subjected. He begins by noting that *all* hypotheses must be sufficient to 'save' the *known* appearances: "the hypotheses which we accept ought [at least] to explain phenomena which we have observed".[15] This, in itself, is a *very* weak test for an hypothesis, and Whewell feels that it is insufficient:

> they [i.e., hypotheses] ought to do more than this: our hypotheses ought to foretell phenomena which have not yet been observed; at least all phenomena of the same kind as those which the hypothesis was invented to explain.[16]

An hypothesis which merely summarizes the known evidence, without successfully going beyond it, scarcely deserves a place in the theoretical structure of science since it merely says, perhaps in different language, what is already contained in the existing observational foundation. However, if the theory can successfully predict results "even of the same kind as those which have been observed, in new cases", then this "is a proof of real success in our inductive process."[17] When a theory successfully manages to make such predictions, we have a natural inclination to regard it as true, or at least very nearly so. This is because the probability of our being able to make successful predictions while using a false hypothesis is on Whewell's view, rather small:

> Men cannot help believing that the laws laid down by discoverers must be in a great measure identical with the real laws of nature when the discoverers thus determine effects beforehand in the same manner in which nature herself determines them ... Those who can do this must to a considerable extent, have detected nature's secret.[18]

But even if successful prediction vastly increases our confidence in a theory, it is still not sufficient to persuade us with certainty of the truth of the theory. Knowing the fallacy of affirming the consequent, and perfectly aware of the correct predictions which false theories had made in the history of science, Whewell is unwilling to take mere predictive success as a sufficient criterion of truth.

However, as one goes further up the scale of increasingly severe tests – to the stage where consilience of inductions begin to occur – it is a very different story. If, instead of being able to predict only phenomena of the same kind as the hypothesis was invented to explain, we can explain and predict, with its help, cases *of a different kind* (relative, of course, to other theories), then we have *indubitable* evidence for the truth of our theory:

These instances in which this [consilience] has occurred, indeed, impress us with a conviction that the truth of our hypothesis is certain. No accident could give rise to such an extraordinary coincidence. No false supposition could, after being adjusted to one class of phenomena, exactly represent a different class, where the agreement was unforeseen and uncontemplated. That rules springing from remote and unconnected quarters should thus leap, to the same point, can only arise from *that* being the point where truth resides.[19]

Notice that this argument is a general one, in the sense that it applies to all three types of consilience ($CI_1 - CI_2$) outlined on page 165. The argument itself clearly begs many questions. The first sentence in the passage quoted above can probably be taken as a reasonably accurate description of the *psychological* force of consiliences. One has only to look at the effect Newton's 'consilience' (the famous 'Newtonian synthesis') had on his followers to see the soundness of Whewell's observation. However, the remainder of this passage seems to be suggesting that this attitude is *logically* justified, that it is simply impossible in principle that any hypothesis could achieve a consilience unless it were the true hypothesis for explaining the phenomena under investigation. But neither in this passage nor elsewhere does Whewell offer any valid argument to support his logical (as opposed to his psychological) claim. Elsewhere in his writings, Whewell similarly asserts the truth-insuring character of consiliences: "When such a convergence of two trains of induction points to the same spot, we can no longer suspect that we are wrong. Such an accumulation of proof really persuades us that we have to do with a *vera causa*".[20] Again he claims – as if repetition were a substitute for justification – that when an

hypothesis, of itself and without adjustment for the purpose, gives us the rule and reason of a class of facts not contemplated in its construction, we have a criterion of its reality, which has never yet been produced in favour of falsehood.[21]

Ultimately Whewell falls back on the old saw (formulated by Descartes,[22] Boyle,[23] and Hartley,[24] among others) that we cannot conceive that a theory which works so successfully in new domains can still be in error. In almost all of his discussions of consilience, Whewell produces a metaphor, probably first elaborated by Descartes in the *Regulae* and in the *Principles,*[25] likening the formation of scientific hypotheses to the decyphering of an inscription in an unknown language. In his reply to Mill, who had denied that consilience is an infallible sign of truth (see below), Whewell formulated the metaphor most explicitly. First, he argues for the strong confirming power of successful *prediction*:

If I copy a long series of letters of which the last half-dozen are concealed, and if I guess those aright, as is found to be the case when they are afterward uncovered, this must be because I have made out the import of the inscription. To say, that because I have copied all that I could see, it is nothing strange that I should guess those which I cannot see, would be absurd, without supposing such a ground for guessing.[26]

Applying this metaphor to the stronger case of consilience, Whewell suggests that we

may compare such occurrences [i.e., consiliences] to a case of interpreting an unknown character, in which two different inscriptions, deciphered by different persons, had given the same alphabet. We should, in such a case, believe with great confidence that the alphabet was the true one.[27]

A consilience thus confers on an hypothesis "a stamp of truth beyond the power of ingenuity to counterfeit".[28]

A second metaphor Whewell sometimes invokes in discussing consilience is the testimony of witnesses to an event. Just as our belief that an event occurred is much stronger if two independent witnesses attest to it than if there is only a single witness, so our confidence in a theory is much greater if it is supported by two separate chains of 'inductive' argument:

It [viz., consilience] is [like] the testimony of two witnesses in behalf of the hypothesis; and in proportion as these two witnesses are separate and independent, the conviction produced by their agreement is more and more complete. When the explanation of two kinds of phenomena, distinct, and not apparently connected, leads us to the same cause, such a coincidence does give a reality to the cause, which it has not while it merely accounts for those appearances which suggested the supposition. This coincidence of propositions is . . . one of the most decisive characteristics of a true theory, . . . [a] Consilience of Inductions.[29]

Whewell maintains that Newton's demand for *verae causae* in his First Rule of Philosophizing was, in reality, a formulation of the requirement of the consilience of inductions.[30] Indeed, the ability of a theory to achieve a consilience is *prima facie* proof that "we have to do with a *vera causa*".[31] He suggests that Newton's *regula prima* can be formulated as the following methodological rule:

we may, provisorily, assume such hypothetical cause as will account for any given class of natural phenomena; but that when two different classes of facts lead us to the same hypothesis, we may hold it to be a *true cause.*[32]

As I said before, Whewell makes a very plausible case for regarding consilience as giving us great confidence in an hypothesis or theory; indeed, a confidence of a much firmer kind than is achieved by the prediction of similars. What his specific arguments obviously fail to establish is that hypotheses which achieve a consilience are true and can be known with certainty to be true. That he is giving an accurate reflection of patterns of belief in the scientific community is beyond doubt. That he has failed to justify such beliefs is equally clear.

In spite of the structural weaknesses in Whewell's argument, his interest in the problem of consilience was of long duration. Not only does it figure in almost all of his published methodological works from 1833 to 1860, but it was even a matter of concern to him in some of his *early* unpublished writings on scientific method. Thus, in a fragment now in the Wren Library of Trinity College, Cambridge, and probably dating from the late 1820s, Whewell formulates something like the requirement of consilience as one of the 'Rules of Philosophizing'. His list of rules is worth reproducing in full:

> Try to put these in a compendious form –
>
> 1. Our hypothesis must be capable of distinctly and appropriately connecting the phenomena.
>
> 2. Generalizations resulting from such hypotheses, if correctly obtained and eminently simple, are theories.
>
> 3. Theories which explain phenomena, detached from those which were used in the generalization, are highly probable, and advance to certainty as the number of unexplained phenomena is diminished.[33]

The third rule is not a straightforward statement of the consilience rule, for it contains no clear requirement that a theory must be able to explain phenomena of a kind *different* from those used to generate the theory in the first place. On its own, therefore, Rule 3 above only makes explicit reference to the fact that theories must be content-increasing. As I mentioned above, this was quite a common requirement by the 1820s. However, in a second draft of those same rules, Whewell adopts a revised form of Rule 3 which moves further in the direction of consilience:

> [Rule] 3. If this theory explains uncontemplated and apparently detached facts it acquires a probability – [and] almost a certainty . . . the theoretical cause becomes a *real cause* (ibid.).

In the revised version, therefore, a theory becomes almost certain if, in addition to explaining facts "detached from those which were used in the generalization" (which was all the first draft required), it explains facts of a kind "uncontemplated" when the theory was discovered. Whewell still has not put the point as succinctly as he will do later. But two things are clear even from this early manuscript: (1) that Whewell was concerned with the problem of consilience at an early stage, and (2) that from the 1820s onwards he was defining Newton's *verae causae* or 'real causes' as those which can achieve a consilience of inductions.

Intertwined in almost all of Whewell's remarks on consilience is his view that scientific laws are necessarily true. Recall that Whewell's 'inductive formula' was a warrant for arguing that if a given hypothesis explains or expresses certain facts, then that hypothesis *may* be the true expression for those facts. Ultimately, however, Whewell wants to establish a much stronger relation. Specifically, he wants the scientist to be able to say that the given facts can be expressed (or explained) *only* by adopting a certain hypothesis.[34] This transition, from viewing an hypothesis as a possible explanation for a set of phenomena to regarding it as the *only* possible explanation for them, is a process which, on Whewell's view, defies logical characterization.[35] It is essentially a temporal process occuring in the mind of a scientist and no formal schema can capture its subtleties. What is clear, Whewell maintains, is that as an hypothesis or lawlike statement is subjected to increasingly difficult trials,

> this conviction, that no other law than those proposed can account for the known facts, finds its place in the mind gradually, as the contemplation of the consequences of the law and the various relations of the facts becomes steady and familiar.[36]

It is really the accretion of a number of different inductive 'proofs' all pointing towards the same conclusion which persuades the scientists that he has discovered a necessary truth. Our *belief* in the truth of an hypothesis becomes so strong that we cannot 'conceive it possible to doubt' the truth of that given hypothesis.[37]

In spite of the fact that Whewell generally regards a successful consilience as an infallible sign of the truth of the theory in which it occurs, he occasionally is less emphatic about identifying consiliatory power with truth. He points out, for example, that the phlogiston theory was capable of explaining facts in such diverse domains as combustion and acidification.[38] Strictly speaking, therefore, the phlogiston theory achieved a 'truth-insuring' consilience of inductions. Nonetheless, as new pheneomena emerged which the phlogiston theory was unable to explain (or able to explain only by *ad hoc* and, for Whewell "inadmissible operations"), that hypothesis was abandoned. Similarly, he concedes (without using this language) that the emission theory of light achieved a consilience in explaining reflection, refraction and (with some difficulty) the colors of thin plates.

John Stuart Mill was quick to see the question-begging character of

Whewell's arguments about the truth-guaranteeing nature of consiliences. In his *System of Logic* (1843), he focussed specifically on Whewell's analogy between establishing the validity of a cypher and proving the truth of an hypothesis. Mill, who held that correct prediction is no more reliable a test of truth than sufficiency to explain the known evidence, argues that:

> If anyone, from examining the greater part of a long inscription, can interpret the characters so that the inscription gives a rational meaning in a known language, there is a strong presumption that his interpretation is correct; but I do not think the presumption much increased by his being able to guess the few remaining letters without seeing them: for we should naturally expect . . . that even an erroneous interpretation which accorded with all the visible parts of the inscription would also accord with the small remainder.[39]

Moreover, as Mill repeatedly stressed, we generally have no guarantee, however much we may have tested our hypothesis, that its next prediction will not be a false one, for there is nothing in principle to prevent an hypothesis which has achieved numerous consiliences from failing on the next occasion when it is put to the test.

Against such an attack, Whewell could reply only by falling back on the rather flimsy empirical observation that theories which have achieved a consilience of inductions have never been subsequently refuted: "the history of science offers no examples in which a theory supported by such consiliences, has been afterwards proved to be false".[40] That Whewell felt he could fall back on the history of science for convincing evidence of the truth-attesting power of consiliences was no accident. On the contrary, the notion of the consilience of induction – though not always by that name – provides the leitmotif for the structure of his *History of the Inductive Sciences* (1837). In the *History* and in the historical Part I of the *Philosophy*, Whewell was concerned to trace the development of science in terms of certain categories of narration which provided natural terminal points for treating the history of a science or a theory. Chief among these narrative categories was his concept of the *inductive epoch*, along with its *prelude* and its *sequel*. Roughly speaking, an inductive epoch is what modern writers would call a scientific revolution. It was characterized by the repudiation of old ideas, the clarification of new conceptions of natural phenomena, and the unification of disconnected facts into a single general theory. The *prelude*, as its name suggests, was the period immediately preceding the inductive epoch, when major problems in the then dominant theory were exposed, and new concepts were articulated. The *sequel*, in its turn, followed the epoch and consisted in the application of theories developed in the epoch to more and more phenomena in an increasingly precise manner. What chiefly characterized the inductive epoch, or scientific revolution, was one or more *major consiliences of inductions.* Thus, Newton's theory achieved a CI_1 *vis-à-vis* the laws of Kepler and Galileo, a CI_2 with respect to the perturbations of the moon, and a CI_3 in its anticipation of the flattening of the earth at the poles. Whewell claims that every scientific revolution is accompanied by, and characterizable in terms of, an associated consilience of inductions:

The great changes which thus take place in the history of science, the revolutions of the intellectual world, have, as a usual and leading character, this, that they are steps of *generalization*; transitions from particular truths to others of a wider extent, in which the former are included.[41]

However, it is not only to static theories or hypotheses that Whewell applies his doctrine of consiliences. He also believes that the consilience notion can function as a device for assessing the cogency and value of prolonged scientific research traditions. If, he argued, we look at the historical development of any scientific school or tradition (e.g., Newtonian mechanics, Cartesian physiology, or catastrophist geology), there are generally clear signs in its development and evolution that its basic theories have become either (1) more complex, artificial and *ad hoc*, or (2) more simplified, more general, more natural, and more coherent. In the former case, the degenerating complication of the theory is probably a sign that it is false, no matter how successful the theory may be in empirically explaining natural phenomena. In the latter case, however, where it has been possible to extend the theory to a wider domain of phenomena with no loss of formal simplicity, we have (Whewell maintains) unambiguous evidence that the theory, at least in its fundamental assumptions, is correct. Clearly case (2), with its content-increasing gains at no loss of formal coherence and unity, is an example of a successful consilience of inductions. Whewell takes the view that it is a defining condition of *progress* in science that 'refuted' theories which are retained in a modified form must ascend to higher and higher levels of generality (by CI_2 or CI_3), whilst converging towards systemic simplicity and unity (by CI_1). As he puts it,

we have to notice a distinction which is found to prevail in the progress of true and false theories. In the former class all the additional suppositions tend to simplicity and harmony; the new suppositions resolve themselves into old ones, or at least require only some easy modification of the hypothesis first assumed: the system becomes more coherent as it is further extended. The elements which we require for explaining a new class of facts are already contained in our system. Different members of the theory run together, and we have thus a constant convergence to unity.[42]

With false theories, on the one hand,

The new suppositions are something altogether in addition; not suggested by the original scheme; perhaps difficult to reconcile with it. Every such addition adds to the complexity of the hypothetical system, which at last becomes unmanageable, and is compelled to surrender its place to some simpler explanation.[43]

On Whewell's view, research traditions are not abandoned merely because they are refuted, but rather because their development does not result in significant consiliences. As he points out in his brilliant "On the Transformation of Hypotheses in the History of Science",[44] virtually any theory can be reconciled with the phenomena if we are prepared to make a sufficient number of *ad hoc* adjustments to it. The only coercive argument against such a patched-up theory cannot be the empirical one that it fails to work, but rather the conceptual one that it has become increasingly complex, cumbersome and logically untidy.

The notion of a consilience of inductions, once formulated by Whewell, became by the late 19th century, a frequently-cited characteristic of sound hypotheses. Some methodologists accepted the requirement of consilience without any reservations. Thus, Stanley Jevons in his *Principles of Science* (1874), echoes Whewell's claim that adequacy to explain the *known* appearances is not enough to make an hypothesis respectable. In what is virtually a paraphrase of Whewell, Jevons goes on to argue:

When once we have obtained a probable hypothesis, we must not rest until we have verified it by comparison with new facts. We must endeavor by deductive reasoning to anticipate such phenomena, especially those of a singular and exceptional nature, as would happen if the hypothesis be true.[45]

In spite of Mill's argument that the successful prediction of novel effects is no guarantee of the truth of an hypothesis, Jevons acquiesces in Whewell's claim that a successful consilience is "the sole and sufficient test of a true hypothesis".[46] Other methodologists, such as Thomas Fowler,[47] conceded that Whewellian consiliences have great psychological impact on the minds of the scientists, but refused to accept Whewell's claim that consilience is a *sufficient* test for truth. Fowler insists that "what is required before a hypthesis can be placed beyond suspicion is *formal proof*" for it, and he charges that Whewell never shows that a "consiliative inference" corresponds to the canons of valid reasoning.[48]

The methodological writings of Peirce are filled with discussions of the problem of consilience, although Peirce rarely calls it by the name. Thus, Peirce is but reformulating Whewell's CI_2, when he writes:

The other variety of the argument from the fulfilment of predictions is where truths ascertained subsequently to the provisional adoption of the hypothesis . . . lead to new predictions being based upon the hypothesis of an entirely different kind from those originally contemplated and these new predictions are equally found to be verified.[49]

And CI_3 is virtually identical to Peirce's demand that a good "hypothesis must be such that it will explain the surprising facts we have before us . . .".[50] In a more recent times, Popper and his school, perhaps more than any other group of contemporary philosophers of science, have addressed themselves to the problem of the consilience of inductions, though they (like Peirce) have not called it by that name. Indeed, Popper's major 'discovery' of the 1950s was a reformulation of the problem of consilience. Until his classic "Three Views Concerning Human Knowledge' (1956), he had not stressed the importance of theories which successfully make novel predictions. Since the mid-fifties, however, Popper, in his many discussions of the criteria for *severe* test, has required a 'good' hypothesis to do precisely what Whewell expected it to do. Consider, for instance, Popper's second "requirement for the growth of knowledge":

. . . we require that the new theory should be *independently* testable. That is to say, apart from explaining all the explicanda which the new theory was designed to explain it must have new and testable consequences (preferably consequences of a *new kind*); it must lead to the prediction of phenomena which have not so far been observed.[51]

Popper's third requirement is that a good theory must also have *passed* "some new and severe tests".[53] Taken together, these requirements correspond almost exactly to Whewell's consilience of inductions, particularly those forms I have identified as CI_2 and CI_3. Popper and Whewell are in full agreement that the best hypothesis or theory is the one which has predicted new phenomena, explained phenomena of different kinds, and made startling predictions. Indeed, the only significant difference between Popper and Whewell on this issue, concerns the degree of confidence to be accorded to an hypothesis which passes severe tests (or, in Whewell's language, which achieves a consilience of inductions).[53]

In spite of the extent to which considerations akin to consilience have found their way into almost every recent account of scientific method, the basic problem to which Whewell addressed himself is still unresolved. Precisely when and how an hypothesis reaches that threshold of confirmation (or severe testing) when it warrants acceptance is as intractable a problem for modern confirmation theorists as it was for Whewell. Like him, they tend to identify that threshold with a successful consilience, or, like Popper, they deny that any such belief-threshold exists. But their justifications for doing so seem no more clear-cut than Whewell's, in spite of the impressive array of formal tools of analysis which they have brought to bear on the problem.

NOTES

[1] Quoted from the *Philosophy of the Inductive Sciences founded upon their History* (2nd ed., London, 1847), Vol. II, p. 469. My italics. Hereinafter cited as '*PIS*'.

[2] For brevity, I shall generally use the term 'consilience' as a shorted form of the phrase 'consilience of inductions'.

[3] "Thus in each inference made by induction, there is introduced some General Conception, which is given, not by the phenomena, but by the mind. The conclusion is not contained in the premises, but includes them by the introduction of a new generality". (Ibid., Vol. II, p. 49.) It is important to stress that the 'new generality' is not merely the generality of the logical operator 'all'; rather, the new generality is a consequence of the fact that "we travel beyond the cases which we have before us; we consider them as exemplifications of some Ideal Case in which the relations are complete and intelligible". (Ibid.)

It was largely Mill's inability to understand this latter sense of 'generality' which provoked the classic Whewell–Mill controversy on induction.

[4] Cf. especially Peirce, *Collected Papers of Charles Sanders Peirce* (Cambridge, Mass., 1934), 5.189, as well as his many other discussions of abduction.

[5] *PIS*, 1847, Vol. II, p. 88.

[6] Throughout Whewell's writings, he frequently refers to different 'types', 'kinds', 'classes' or 'domains' of facts. However, he never concerns himself with providing criteria for determining that facts are of similar or different kinds. Whewell, however, is not alone in this, for more recent writers like Carnap (see for instance, Carnap's *Logical Foundations of Probability* (Chicago, 1962), p. 575) and Popper (note 5, below), who adopt principles similar to (1) and (3), are equally vague in referring to events of 'different kinds'. A similar ambiguity can be found in most discussions of the so-called Principle of Limited Variety.

[7] *PIS*, 1847, Vol. II, p. 65. Elsewhere he describes this species of consilience in the following terms: "if we again find other facts of a sort uncontemplated in framing our hypothesis, but yet clearly accounted for when we have adopted the supposition; – we are ... [led] to believe [it] without conceiving it possible to doubt" (Ibid., Vol. II, p. 286. Cf. also ibid., Vol. II, pp. 427–28.)

[8] In general, Whewell commentators and scholars have not seen that his notion of consilience of inductions embraces these different characteristics. Robert Butts, for instance, once wrote as if Whewell's theory of consilience requires no more than that a theory should successfully explain more than it was originally devised to explain. (See Butts, *William Whewell's Theory of Scientific Method* (Pittsburgh, 1968), p. 18). As I point out below, and as Butts himself has subsequently stressed (see especially his contribution to Giere and Westfall (eds.), *Foundations of Scientific Method in the 19th Century* (Bloomington, 1973) this is a misreading of Whewellian consilience. On Butts' original account, consilience amounts to the extremely weak requirement that a theory

should go *beyond* its initial evidence. As other chapters in this book make clear, this doctrine has been a commonplace since the time of Bacon, and certainly represents no new insight on Whewell's part. If, on the other hand, my interpretation of Whewellian consilience is more or less correct, then Whewell's demand for consilience is a much more subtle and exacting requirement than was formerly thought.

In his otherwise excellent study *Le rationalisme de Whewell* (Paris, 1935), Robert Blanché devotes less than a page to the consilience of inductions. C. J. Ducasse similarly oversimplifies the matter by suggesting that the requirement of consilience is "in its essence identical [sic] with ... the well-known maxim connected with the name of William of Occam" (Blake, Ducasse & Madden, *Theories of Scientific Method: The Renaissance through the Nineteenth Century* (Seattle), 1960), p. 212).

[9] See, for instance, Butts, op. cit., note 8.

[10] But not always, for an hypothesis may explain some very surprising phenomenon which was *already known* when the hypothesis was devised.

[11] Note that CI_2 is to be distinguished sharply from the case where an hypothesis or theory merely predicts events of a kind *similar* to those already known when the theory was conceived. In the latter case (which I discuss below) no consilience of inductions occurs, even though our adoption of the hypothesis is content increasing.

[12] *PIS*, 1847, Vol. II, pp. 67–8. Cf. his remark that "when such a convergence of two trains of induction points to the same spot, we can no longer suspect that we are wrong" (Ibid., Vol. II, p. 286).

[13] To my knowledge, the only philosopher of science before Whewell to formulate a requirement like CI_3 was John Herschel. In his *Preliminary Discourse on the Study of Natural Philosophy* (London, 1830), he observed that "the surest and best characteristic of a well-founded and extensive induction ... is when verifications of it spring up, as it were, spontaneously into notice, from quarters where they might be least expected, or even among instances of that kind which were at first considered hostile to them. Evidence of this kind is irresistible and compels assent with a weight which scarcely any other possesses" (op. cit., para 180). Nonetheless, there is a subtle but important difference between Herschel's requirement and Whewell's CI_3. Whereas Whewell attaches greatest importance to the explanation of *surprising* facts, Herschel seems to lay greatest stress on the successful explanation of facts which had previously been regarded *as counter-examples*.

[14] For the evidence for this claim, see Chapters 4, 5, and 11.

[15] *PIS*, 1847, Vol. II, p. 62.

[16] Ibid., pp. 62–3.

[17] Ibid., p. 152.

[18] Ibid., p. 64.

[19] Ibid., p. 65. In a similar vein, he observes that "when the explanation of two kinds of phenomena, distinct, and not apparently connected, leads us to the same cause, such a coincidence does give a reality to the cause, which it has not while it merely accounts for those appearances which suggested the supposition" (ibid., p. 285).

[20] Ibid., p. 285–6.

[21] Ibid., pp. 67–8.

[22] "Although there exist several individual effects to which it is easy to adjust diverse causes [i.e., hypotheses], one to each, it is however not so easy to adjust one and the same [hypothesis] to several different effects, unless it be the true one from which they

proceed" (Descartes, *Oeuvres*, ed. Adam & Tannery (Paris, 1897–1957), Vol. II, p. 199. Cf. also ibid., Vol. IX, p. 123).

23 "For it is much more difficult to find an hypothesis, that is not true, which will suit with many phenomena, *especially if they be of various kinds*, than but with a few" (Boyle, *Works*, ed. Birch (London, 1772), Vol. IV, p. 234). For a detailed discussion of Boyle's methodology, see Chapter 4.

24 "And if the general conclusion or law be simple, and always the same, from whatever phenomena it be deduced ·. . . there can scarce remain any doubt, but that we are in possession of the true law inquired after . . . " (Hartley, *Observations on Man* (London, 1791), Part I, p. 341).

25 This analogy has a long pedigree. Descartes introduced it into modern philosophy in his *regula* X and in *Oeuvres*, Vol. IX, p. 323. As I showed in Chapter 4, Boyle, Glanvill, Power and Locke borrowed the analogy from Descartes. For Hartley's version of the metaphor, see Chapter 8. See also, for other discussions of this methodological metaphor, Boyle's *Of the Excellency and Grounds of the Corpuscular or Mechanical Philosophy*; Leibniz's Letter to Conring; his *Elementa Physicae* and his *New Essays*, Chap. XII; Boscovich's *De Solis ac Lunae Defectibus*, p. 211–12; 'sGravesande's *Introductio ad Philosophiam*; D'Alembert's article 'Dechiffrer' for the *Encyclopédie*; and Dugald Stewart's *Elements of the Philosophy of the Human Mind.*

26 *On the Philosophy of Discovery* (London, 1860), p. 274. (Hereinafter cited as '*POD*'.)

27 Ibid., pp. 274–5.

28 Ibid., p. 274.

29 *PIS*, 1847, Vol. II, p. 285.

30 Robert Butts has given an extremely valuable exegesis of Whewell's attempt to interpret Newton's doctrine of *verae causae* in his "Whewell on Newton's Rules of Philosophizing", in Butts & Davis, (eds.), *The Methodological Heritage of Newton* (Toronto, 1970).

31 *PIS*, 1847, Vol. II, p. 286.

32 Ibid. As he says elsewhere, "Newton's [First] Rule of Philosophizing will become a valuable guide, if we understand it as asserting that when the explanation of two or more different kinds of phenomena . . . lead[s] us *to the same* cause, such a coincidence gives a reality to the cause. We have, in fact, in such a case, a Consilience of Inductions". *POD*, pp. 276–7).

33 Trinity College Add. MS. a 78[60]. The manuscript is undated, and my dating is based chiefly on handwriting patterns. It is quoted with the kind permission of the Master and Fellows of Trinity College.

34 The ideal inductive formula, in Whewell's view, is this: "The several facts are exactly expressed as one fact if, *and only if*, we adopt the Conception and the Assertion of the inductive inference" (*PIS*, 1847, Vol. II, p. 90).

35 As he says mysteriously in *On the Philosophy of Discovery*: "Induction is inconclusive *as reasoning*. It is not reasoning: it is another way of getting at truth".

36 *PIS*, 1847, Vol. II, p. 622.

37 Ibid., Vol. II, p. 286. For a lucid account of Whewell's theory of necessary ideas, see Robert Butts, 'Necessary Truth in Whewell's Theory of Science', *American Philosophical Quarterly* 2 (1965), 1–21. I am in sufficiently close agreement with Butts' analysis of Whewellian necessity that I do not propose to discuss it further here.

[38] *PIS*, 1847, Vol. II, pp. 70–71.
[39] J. S. Mill, *System of Logic* (8th ed., 1961), p. 329.
[40] *POD*, p. 275. Elsewhere, he claims that "there are no instances in which a doctrine recommended in this manner [i.e., by a consilience of inductions] has afterwards been discovered to be false". (*PIS*, Vol. II, p. 286). And again: "No example can be pointed out, in the whole history of science so far as I am aware, in which this Consilience of Inductions has given testimony in favour of an hypothesis afterwards discovered to be false" (Ibid., Vol. II, p. 67).
[41] *History of the Inductive Sciences* (3d ed., London, 1857), Vol. I, p. 46.
[42] *PIS*, 1847, Vol. II, p. 68.
[43] Ibid., pp. 68–9.
[44] *Transactions of the Cambridge Philosophical Society* 9 (1851), 139–47.
[45] Jevons, *The Principles of Science* (2d ed; London, 1924), p. 504. See also Ernst Apelt's *Die Theorie der Induction* (Leipzig, 1854), 186–7, and J. Stallo's *Concepts and Theories of Modern Physics* (Cambridge, Mass, 1960), p. 136.
[46] Jevons, *Principles of Science*, loc. cit.
[47] Cf. Fowler's *Elements of Inductive Logic* (Oxford, 1872), pp. 112–14.
[48] Ibid., p. 114.
[49] *Collected Papers of Charles Sanders Peirce* (Cambridge, Mass, 1958), 7.117.
[50] Ibid., 7.220. See also ibid., 7.58 and 7.115–16. Any number of similar formulations could be gleaned from a careful pruning of Peirce's writings.
[51] Popper, *Conjectures and Refutations* (N. Y., 1968), p. 241. Italics in original. In his "Three Views . . . ", Popper insists that "there is an important distinction . . . between the prediction of *events of a kind which is known*, . . . and, on the other hand, the prediction of *new kinds of events* (which the physicist calls 'new effects')" (op. cit., p. 117).
[52] Ibid., p. 242.
[53] Another important difference is that Popper curiously denies that consiliences of type CI_1 are of any methodological significance whatever.

CHAPTER 11

WHY WAS THE LOGIC OF DISCOVERY ABANDONED?

It is difficult to find a problem area in the philosophy of science about which more rubbish has been talked and in which more confusion reigns than 'the philosophy of discovery'. It is even hard to keep the characters straight. Russ Hanson, who thought the logic of discovery was a good thing, advocated the method of abduction, which was a method for the evaluation, not the discovery, of hypotheses. Hans Reichenbach, who was notorious for insisting that the 'context of discovery' is of no philosophical significance, was a proponent of the straight rule of induction, a technique for the discovery of natural regularities if ever there was one. Not to be slighted here is Karl Popper who wrote a book called the *Logic of Scientific Discovery*, which denies the existence of any referent for its title.

In the circumstances, it will perhaps not be taken amiss if I suggest that some historical explorations may be appropriate. With any luck, an analysis of the history of ideas about the logic of discovery may put us in a position to see what significance, if any, should be attached to recent efforts to get clear about the philosophical problems of discovery.

Not the least of my concerns will be that of utilizing history to ascertain what problems, what important questions, a logic of discovery was thought to resolve. If the recently revived program for finding a logic of discovery is worth taking seriously, it must be because such a logic, once propounded, might illuminate or clarify some important philosophical problems about the nature of science. Perhaps by examining the historical development of that tradition, we can tell what problems lie behind the quest for a logic of discovery.

Before we get to the history itself, a clarificatory preamble is in order. The term 'logic of discovery', like 'discovery' itself, is notoriously ambiguous. If one views the logic of justification as concerned exclusively with a study of the evidence relevant to the proverbial 'finished research report', then the logic of discovery – construed as dealing with the development and articulation of an idea at every stage in its history prior to its ultimate ratification – has a very wide scope indeed. It would include an account of how a theory was first invented, how it was preliminarily evaluated and tested, how it was modified, and the like. I do not intend to interpret discovery so broadly.

Between the context of discovery and the context of ultimate justification, there is a nether region, which I have called the *context of pursuit.*[1]

In the context of pursuit, the constraints are (presumably) tighter than the constraints (if any there be) on discovery and significantly looser than those we insist upon where belief or acceptance is concerned. This 'three-fold way' has several virtues lacking in the usual dichotomy. For one thing, these three contexts mark the temporal, if not the logical, history of a concept. It is first discovered; if found worthy of pursuit, it is entertained; if further evaluation shows it to be worthy of belief, it is accepted. More importantly, this trichotomy prevents us from lumping together activities and modalities of appraisal which have frequently been confused with one another. For instance, both Peirce and Hanson construed the method of 'abduction' as a technique of scientific discovery. But it is nothing of the kind. As Wesley Salmon and others have pointed out, abduction does *not* tell us how to invent or discover an hypothesis. It leaves that (possibly creative) process unanalyzed and tells us instead when an idea is worthy of pursuit (namely, when it explains something we are curious about). By calling the abduction method a partial logic of discovery, Hanson and to a lesser degree Peirce are culpable of having obfuscated the real nature of a logic of discovery. Equally, many of Hanson's critics have erroneously concluded that because abduction is no method of discovery, it must necessarily belong to the context of justification. Neither party to the dispute has seen that the abductive method belongs most naturally to neither discovery nor justification, but rather to pursuit.

However, my purpose in the early part of this chapter is not to engage in a contemporary philosophical debate. The aim, rather, is to look at the historical development of views about the logic of discovery in order to ascertain why the optimism and the urgency among our philosophical forebears about understanding the process of discovery has generally given way to pessimism about the very possibility of a *philosophical* account of discovery.

These contentious preliminaries will have served their purpose, however, if they permit me to say that my concerns here will be with views about scientific discovery, not scientific pursuit. Accordingly, I shall construe discovery rather narrowly as concerned with 'the *eureka* moment', i.e., the time when a new idea or conception first dawns, and I shall view the logic of discovery as a set of rules or principles according to which new discoveries can be *generated.* The historical justification for such a construal is that this latter sense is precisely what was traditionally meant by the logic or philosophy or 'art' of discovery. The philosophical rationale for interpreting the logic of discovery in its narrow, rather than its broad sense is that *only* on this construal can any sense be given to the current debate about the existence of a logic of discovery; since I take it that no one would deny that there clearly are rules or general principles governing discovery, if that term is construed so broadly as to include pursuit.

An event of major significance occurred in the course of 19th-century philosophy of science. The task of articulating a logic of scientific discovery and concept formation – a task which had been at the core of epistemology since Aristotle's *Posterior Analytics* – was abandoned. In its place was put the very different job of formulating a logic of *post hoc* theory evaluation, a logic which did not concern itself with how concepts were generated or how theories were first formulated. This transformation marks one of the central watersheds in the history of philosophical thought, a fundamental cleavage between two very different perspectives on how knowledge is to be legitimated.

Throughout the 17th and 18th centuries (as in antiquity), the enterprise of articulating an organon of discovery, an *ars inveniendi*, flourished. Bacon, Descartes, Boyle, Locke, Leibniz, and Newton – to name only the more prominent – all believed it was possible to formulate rules which would lead to the discovery of 'useful' facts and theories about nature. By the last half of the 19th century, this enterprise was dead, unambiguously repudiated by such philosophers of science as Peirce, Jevons, Mach, and Duhem. This chapter is a speculative attempt to explain the changing historical fortunes of the logic of discovery; speculative, because the space available to me does not allow for the sort of detailed historical documentation which would be required to make the story convincing. I shall aim, more modestly, at making the story at least plausible.

The desire to develop a logic of discovery was generally based upon two quite different motives. On the one hand, there was the heuristic and pragmatic problem of how to accelerate the pace of scientific advances, of how to increase the rate at which new discoveries were made by articulating fruitful rules for invention and innovation. (This side of the issue was especially prominent in Bacon, although there are clear signs of it in Descartes and Leibniz as well.) On the other hand, and more importantly from a philosophical point of view, there was the *epistemological* problem of how to provide a sound warrant for our claims about the world. If a fool-proof logic of discovery could be devised, it would solve both problems simultaneously. It would both be an *instrumentarium* for generating new theories and, because infallible, it would automatically guarantee that any theories produced by it were epistemically well grounded.

This second function is the crucial one to stress here. As conceived by most 17th- and 18th-century authors, *a logic of discovery would function epistemically as a logic of justification*. Unlike now, there was then no distinction drawn between the contexts of discovery and justification. It is not that our forebears could not recognize the difference between discovery and justification; they could and often did. But they were convinced that an appropriate (i.e., infallible) logic of discovery would automatically authenticate its products and that a separate logic of justification would therefore be redundant and unnecessary. They were preoccupied with developing logics of discovery, not because they were indifferent to the epistemological problem of justifying knowledge claims, but precisely because they took the justificational problem to be central.

Let me unpack this a bit more. Simply put, thinkers since antiquity had explored three rival views about how to justify the claim of natural philosophy to be a genuine *scientia*. On one possible account – which corresponds to a familiar 20th-century caricature of Descartes – scientific claims were self-authenticating. They had only to be understood to see that they were true and could not be otherwise. There is scarcely any major figure in the history of philosophy, including Descartes, who maintained that most scientific theories could be authenticated in this way. Virtually all writers agreed that scientific truths about the world could only be 'educed' from contingent empirical data about the world; they agreed that there was no warranting process for typical scientific theories which did not involve *a posteriori* evidence. But beyond this point of agreement, there was a significant parting of the ways. One group – whom I shall call the consequentialists – believed that theories or claims could be justified by comparing (a subset of) their consequences with observation. If an appropriately selected range of consequences proved to be true, this was thought to provide an epistemic justification for asserting the truth of the theory. A second, and more predominate group, whom I shall call the *generators*, believed that theories could be established only by showing that they followed logically (using certain allegedly truth-preserving algorithms) from statements which were directly gleaned from observation.

Much of this chapter will be about the debate between the consequentialists (who tended to stress the method of hypothesis and *post hoc* confirmation) and the generators (who believed that generational algorithms were the ideal device for authentication). What needs to be stressed at the outset is that *both groups* – including the latter, who argued for a logic of discovery – *were primarily concerned with the epistemic problem of theory justification*.

The historical vicissitudes of the generators' program for establishing a logic of discovery are utterly unintelligible, I submit, unless one realizes that the *raison d'être* for seeking a logic of discovery was chiefly to provide a legitimate logic of justification.

As Dewey and Popper have stressed, much of the history of epistemology is characterized by an infallibilist bent. Aristotle and Plato, Locke and Leibniz, Descartes and Kant: all subscribed to the view that legitimate science consists of statements which are both true and known to be true. To provide an epistemic justification for an assertion, i.e., to show that it is 'scientific', was to point to evidence and rules of inference which, collectively, guaranteed its truth. We shall shortly come to see that this infallibilist orientation was closely linked with the generators and the logic of discovery.

Among both ancient astronomers and ancient physicians, consequentialism was a prevalent doctrine. The Hippocratic tradition and the saving-the-phenomena tradition advocated the view that theories were to be judged in terms of an assessment of the truth of their consequences.[2] Eschewing the belief that theories could somehow be derived from observation, these thinkers stressed the necessity for the *post hoc* evaluation of a proposed theory in light of how well it stood up to observational and experiential tests.

But Plato and Aristotle, thorough-going infallibilists where *episteme* was concerned, quickly saw the flaw in consequentialism. To argue from the truth of a consequence of a theory to the truth of the theory itself is, as they perceived, a logical fallacy (the so-called fallacy of affirming the consequent). They quite correctly argued that an infallibilist view of science was strictly incompatible with any form of consequentialism, i.e., with any form of *post hoc* empirical testing, since the latter was logically inconclusive. If theories were to be demonstrably true, such demonstration could not come from any (nonexhaustive) survey of the truth of their consequences.

Because consequentialism was generally acknowledged to be epistemically inconclusive, most of those writers who believed science was infallible knowledge opted for some form of generationism. After all, if truth-preserving rules could be found for moving from particulars to universals, *and* if reliable particulars could be got at (which few seriously doubted), then one would have all that was required to give a justification of science as infallible knowledge. The point is this: if one seeks infallible knowledge and if one grants the fallaciousness of affirming the consequent, then the only viable hope for a logic of justification will reside in the quest for a truth-preserving logic of discovery. Except in very special circumstances,[3] infallibilism leads ineluctably to generationism (although *not* vice versa) and to the obliteration of any significant distinction between the contexts of discovery and justification.

A part, then, of the attractiveness of a logic of discovery was linked to the prevalence of an infallibilist view of science. So long as the latter was in fashion, the former would be likely to remain in vogue. But this linkage is only one part of a very complex story. Another piece of the puzzle is closely related to views about the *aims* of science.

I can motivate this linkage by beginning with an observation on our own time. If there is general scepticism today about the viability of a logic of discovery, it is in part because most of us cannot conceive that there might be rules that would lead us from laboratory data to theories as complex as quantum theory, general relativity, and the structure of DNA. Our shared archetypes of significant science virtually all involve theoretical entities and processes which are inferentially far removed from the data which they explain. That there might be rules to lead one from tracks on a photographic plate to claims about the fine structure of subatomic particles is, to say the least, implausible. But suppose our archetypes were rather different; suppose specifically that they were such lawlike statements as "All crows are black", "All gases expand when heated", or "All planets move in ellipses". If we viewed these as the important products of scientific inquiry, the notion that there might be some algorithms for going from evidence to discovered theory would not seem so bizarre. (Indeed, induction by simple enumeration, Peirce's 'qualitative induction' and Reichenbach's version of the straight rule are precisely such algorithms.)

The point is this: if we expect science to consist chiefly of general statements concerning observable regularities, then mechanical rules for generating a universal from one of more of its singular instances are not out of the question. By contrast, if we expect science to deal with 'deep-structure', explanatory theories (some of whose central concepts have no observational analogues), then the existence of plausible rules of discovery seems much more doubtful.

The historical significance of this contrast is that we should expect to find the search for logics of discovery much more fashionable in epochs when the goal of science is seen as discovering empirical laws than in epochs when the stress is upon discovering explanatory, deep-structural theories. This conjecture is generally confirmed by the historical record. Thus, during much of the 18th century, when Bacon and Newton had persuaded most philosophers that speculative theories and unobservable entities were anathema, inductive logics of discovery were ubiquitous among empiricists. Hume, Reid, D'Alembert, Priestley, and a multitude of other enlightenment philosophers took for granted the existence of an inductive engine of discovery. Similarly in the 1930s and 1940s, when again philosophy was dominated by the Spartan view that science consists chiefly of inductive generalizations, thinkers like Reichenbach, Cohen, and Nagel could devote much energy to discussing various inductive rules of discovery.

By contrast, in epochs when empirical generalizations are viewed as mundane and where theories are conceived chiefly as grandiose ontological frameworks, replete with unobservable entities, inductive logics of discovery have been ignored or, in some cases, their very existence denied. This was true of many 19th-century thinkers (such as Whewell and Boltzmann) who were struck by the nonobservational character of many of the chief theories of their time (especially important here were the wave theory of light and the atomic theory). It also seems to be true of our own time, when few mainstream philosophers show much interest in the discovery of empirical regularities.

But the historical record also establishes that our *a priori* analysis was over-hasty in one respect. While it is true that logics of discovery which involve some form of enumerative induction are taken seriously only by philosophers who believe that 'observational laws' typify scientific inquiry, there are other, non-inductive logics of discovery which have been invoked by those concerned with the analysis of full-blown theories. Indeed, if we look carefully at the writings of many 18th- and 19th-century philosophers of science (including Hartley, LeSage, Priestley, and Peirce),[4] we frequently find a concern with modes or methods of discovery which are quite unlike enumerative induction. I shall call these 'self-corrective logics of discovery'. Such 'logics' involve the application of an algorithm to a complex conjunction which consists of a predecessor theory and a relevant observation (usually one that refutes the prior theory). The algorithm is designed to produce a new theory which is 'truer' than the old. Such logics were thought to be analogous to various self-corrective methods of approximation in mathematics, where an initial posit or hypothesis was successively modified so as to produce revised posits which were demonstrably closer to the true value.

The appeal of this form of the logic of discovery was that, unlike inductive logics of discovery, it did not restrict itself exclusively to statements about observables. The avowed aim, in exploring such a logic of discovery,was to have a logic rich enough that it could deal with the genesis of deep-structural theories. Unfortunately, as I show in Chapter 14, a century-and-a-half of exploration by a succession of major thinkers failed to bring the self-corrective program to fruition. (Peirce, for instance, grappled with this program for forty years before abandoning it.) No one was able to suggest plausible rules for modifying earlier theories in the face of new evidence so as to produce demonstrably superior replacements.

Thus far I have not been offering history, but rather some conceptual preliminaries with which I hope to make the history intelligible. The historical story itself is easy to tell. Most 17th- and 18th-century writers were epistemic infallibilists; accordingly, they looked chiefly to a logic of discovery rather than a logic of testing to provide the indubitable warrant for genuine science. Among the empiricists of the period (especially Locke, Newton, and Hume), there was the added conviction that laws linking observables – rather than 'transductive' theories explaining observables – were the hallmark of genuine science. This, too, conduced to make a justificatory and inductive logic of discovery plausible.

By the 1750s, the picture becomes murkier. As I showed in Chapter 8, explanatory theories became fashionable again (subtle fluids, atoms, aethers, etc.). Accordingly, there was a distinct shift away from 'inductive' logics of discovery towards self-corrective logics of theoretical discovery (these are prominent in the writings of Hartley, LeSage, and Priestley). Infallibilism, however, remained the epistemic orthodoxy.

By the 1820s and 1830s, infallibilism itself was crumbling. Herschel, Whewell and Comte all acknowledged that there is no formula for producing true theories. As fallibilism emerged, there was an unmistakable shift away from the analysis of genesis towards the *post hoc* evaluation of theories. It was argued that theories could not be proven to be true and that the most we can expect is that they can be shown to be likely or probable. The task of justification within such an orientation becomes the more modest (if still troublesome) one of showing that theories are likely. As soon as epistemic fallibilism replaces infallibilism, the possibility of justifying theories by examining their consequences becomes viable again; for all the familiar arguments about the impotence of *post hoc* confirmation to *prove* theories cease to be relevant once justification is no longer perceived as a matter of proof.

Thus, nongenerational logics of justification and epistemic fallibilism emerge simultaneously in the 19th century and jointly render redundant any epistemic role for the logic of discovery. The confluence of these ideas can be shown clearly in the works of Herschel and Whewell, who were among the first philosophers of science to stress that theories could be judged independently of a knowledge of their mode of generation.

Herschel puts the point succinctly:

In the study of nature, we must not, therefore, be scrupulous as to *how* we reach to a knowledge of such general facts, i.e., laws and theories: provided only we verify them carefully when once discovered, we must be content to seize them wherever they are to be found.[5]

Herschel's point involves no denial of the existence of a logic of discovery; he is rather asserting the *irrelevance* of the manner of generation of a hypothesis to its evaluation or justification. Whewell goes a step farther and denies the very existence of a logic of scientific discovery:

Scientific discovery must ever depend upon some happy thought, of which we cannot trace the origin; some fortunate cast of intellect, rising above all rules. No maxims can be given which inevitably lead to discovery.[6]

What Herschel and Whewell are pointing out (and their contemporary Comte makes a similar argument) is that (1) theories can be appraised ('verified') independently of the circumstances of their generation, and (2) such modes of appraisal, even if fallible, are more germane to the process of justification than any fallible rules of discovery would be. Where earlier philosophers had believe that it was only *via* a logic of discovery that theories could be justified, Herschel, Comte, and Whewell sever that link, insisting that justification need not be parasitic on the manner of generation.

This shift in sensibilities is, if anything, even more drastic than I have described it to be. Among the 'generators', the crucial evidence for a theory – indeed the *only* relevant evidence for a theory – consisted in those observations which were temporally (and thus epistemically) prior to the theory itself. It was, after all, such data from which the theory was to be educed. But once Herschel, Whewell, and Comte get their hands on methodology, they radically invert all these relationships. As I showed in Chapter 8, they stress that the strongest evidence for a theory is precisely that evidence which was *not* available when the theory was devised and thus could not conceivably have been used in its genesis. They tended heavily to discount the evidential significance of data available to inventors of the theory. So, it is not merely that the 'new wave' methodologists argued that it was a matter of indifference whence a hypothesis or theory first arose; *they positively insisted that data utilized in generating a theory – which had been the unique form of evidence in the logic of discovery – was to be given less evidential weight than data which were altogether unknown in the context of discovery*. Whewell puts it in a nutshell when he says that a theory acquires some plausibility "by its fully explaining what it was first meant to explain", but it is only properly "confirmed" "by its explaining what it was *not* meant to explain".[7]

The divorce of justification from discovery urged by methodologists in the 1830s and 1840s quickly became the philosophical orthodoxy. The lack of serious resistance to that separation is understandable, *provided* one realizes that justification had been the central problem all along. As soon as justification was separated from proof and as soon as plausible modes of justification were suggested which circumvented the context of discovery, most philosophers of science willingly replaced the program for finding a logic of discovery – which was foundering anyway for the reasons mentioned above – by a program for defining post-discovery empirical support.

This latter program, of course, has characterized most of the philosophy of science since the middle of the 19th century. Some recent writers who would revive an interest in the logic of discovery, see it as something very different from the logic of justification. In this sense, they are radically at odds with the traditional aims of the logic of discovery. The older program for a logic of discovery at least had a clear philosophic rationale: it was addressed to the unquestionably important philosophical problem of providing an epistemic warrant for accepting scientific theories. The newer program for the logic of discovery, by contrast, has yet to make clear what philosophical problems about science it is addressing.

I have sought to show in this chapter that the primary motivation for seeking a logic of discovery was to exhibit the well-foundedness of knowledge claims about the world. The program for articulating an infallible logic of discovery never came to fruition; but that failure only partially explains its abandonment. Equally crucial here was the joint emergence of epistemic fallibilism and of *post hoc* logics of theory testing; developments which rendered redundant and supernumerary the logic of discovery so far as the epistemic issue is concerned. It remains redundant now.

At the same time, it should be stressed that there are also several important *heuristic* problems about science: How can we maximize the rate at which new and promising theories and laws are generated? Once a theory is refuted, in what ways can it be modified to accommodate the recalcitrant data? *Post hoc* logics of justification have nothing to contribute here and can in nowise fulfill the heuristic tasks conceived for a logic of discovery. Nor can such logics of justification fulfill any of the other tasks associated with that part of methodology called heuristics. But before one concludes that the logic of discovery still has a philosophical rationale, one must ask what is specifically *philosophical* about studying the genesis of theories. Simply put, a theory is an artifact, fashioned perhaps by certain tools (e.g., implicit rules of 'search'). The investigation of the mode of manufacture of artifacts (whether clay pots, surgical scalpels, or vitamin pills) is not normally viewed as a philosophical activity. And quite rightly, for the techniques appropriate to such investigations are those of the empirical sciences, such as psychology, anthropology, and physiology. The philosopher of art is not concerned *qua* philosopher with how a sculpture is chiseled out of a piece of granite; nor is the philosopher of law concerned with the mechanics of drafting a piece of legislation. Similarly, the case has yet to be made that the rules governing the techniques whereby theories are invented (if any such rules there be) are the sorts of things that philosophers should claim any interest in or competence at. If this chapter provides a partial answer to the question "Why was the logic of discovery abandoned?", it poses afresh the challenge: "Why should the logic of discovery be revived?" It may be possible to give a convincing and an affirmative answer to the latter question, and much of the recent work of Thomas Nickles promisingly stakes out the relevant territory.[8] But in whatever guise the logic of discovery may eventually appear, it seems clear that its traditional *epistemic* role has been well and truly pre-empted by the theory of testing.

NOTES

1 For a discussion of this problem, see my *Progress and Its Problems* (Berkeley, 1977), pp. 108–14.

2 See especially P. Duhem, *To Save the Phenomena* (Chicago, 1973).

3 The circumstances to which I refer are of two kinds: (a) if one believed that all the consequences of a theory could be examined, then consequentialism and infallibilism would be compatible; (b) if one believed that all possible theories could be enumerated and rejected seriatim by a method of exhaustion, then consequentialism and infallibilism are also compatible. For those many writers who rejected both of these assumptions, infallibilism ruled out any form of consequentialism.

4 For lengthy discussions of the views of these writers on the methodology of science, see Chapter 8 and 14.

5 J. F. W. Herschel, *Preliminary Discourse on the Study of Natural Philosophy* (London, 1830), p. 164.

6 W. Whewell, *Philosophy of the Inductive Sciences*, 2nd ed. (London, 1847), vl. II, pp. 20–21.

7 W. Whewell, *History of the Inductive Sciences*, 3rd ed. (London, 1857), vol. II, p. 370.

8 See especially Nickles' introductory essay in T. Nickles (ed.), *Scientific Discovery, Logic and Rationality* (Dordrecht, 1980).

CHAPTER 12

A NOTE ON INDUCTION AND PROBABILITY IN THE 19TH CENTURY[1]

This short chapter addresses itself to two of the more puzzling features of the historical development of the philosophy of science; first, why did it take so long for philosophers of science to bring the techniques of the mathematical theory of probability to bear on the logic of scientific inference? Why did we have to wait for Stanley Jevons, and C. S. Peirce, writing in the 1870s, rather than Hume in the 1740s or Mill in the 1840s, to find someone systematically arguing that inductive logic is based on probability theory? Still more curious is the fact that Jevons' well-publicized attempt to 'reduce' induction to probability was repudiated by the great majority of the 'inductive logicians' of his day (Peirce being the major exception), and that his program lay dormant for half a century until it was resurrected by Keynes, Carnap and Reichenbach in the 1920s and 1930s. What sort of obstacles made it impossible for our forebears to accept what many philosophers of science today view as the self-evident commonplace that probability theory provides the language for inductive logic? In this chapter, I want to suggest some tentative answers to these very complex historical puzzles.

The first of our two puzzles is easier to resolve than the second. The application of probability to induction was not taken seriously until the 19th century, largely because there was thought to be no significant element of uncertainty or doubt attached to the conclusions of so-called inductive inference. Virtually every major theory of induction developed before the middle of the 19th century was supposed to guarantee the truth of the conclusions to which its application led. Thinkers as diverse as Bacon, Hooke, Lambert, Herschel, Newton, Whewell, and Mill had argued for various forms of inductive inquiry which were all allegedly infallible. Although all these authors realized the inconclusiveness of induction by simple enumeration, they believed that other inductivist strategies existed which would ineluctably lead to true conclusions. Thus, Mill's four canons of inductive inquiry, Bacon's tables, Whewell's method of the consilience of inductions, and Herschel's method of 'first-order inductions' were all designed as infallible tests of putative theories.

In an intellectual atmosphere optimistic about the possibility of an infallible empirical science, it is hardly surprising that it occurred to few people to apply techniques of probability theory to the analysis of scientific inference. Since the conclusions of scientific inferences were thought to be certain rather than probable, it would have seemed an epistemic category mistake to introduce probabilistic techniques into a theory of scientific inference.

As a result of developments which I cannot discuss here, each of these inductive logics was eventually discovered to have no built-in device for guaranteeing the truth of the propositions it produced. Moreover, each was shown to rest on assumptions (in Mill's case, for instance, the principle of limited variety) which were highly controversial. As a result of a growing recognition of the limitations to all known inductive logics (interpreted as instruments of proof), many methodologists around the middle of the last century began to suggest that the conclusions of scientific inference were not certain, but merely probable. Inductive logic came increasingly to be seen as a tool for assessing the relative likelihood of competing scientific theories and hypotheses. Methodologists began to talk about scientific theories as more or less probable, more or less worthy of rational belief. This movement was further stimulated by the increasing importance of probabilistic techniques in the natural sciences. Bunsen and Kirchoff had used statistical methods to analyze the elementary composition of the solar spectrum; Herschel's study of crystalline polarity and Michell's earlier work on double stars similarly served to direct the attention of philosophers of science to problems of statistical inference.

It was in this general context that philosophers first began seriously to suggest that the mathematical theory of probability might have something significant to contribute to the logic of science. After all, probability theorists – especially Laplace – had been writing for over a century about the probability of causes. Although this literature had been largely ignored by inductive logicians in the first half of the 19th century, the crisis provoked by the abandonment of classical theories of induction made it natural for philosophers of science to wonder if the resolution of their dilemma might not lie in the mathematical theory of probability, particularly that portion of it known as the theory of inverse probability. Specifically, people began to try to reduce inductive logic to inverse probability theory.

Whatever specific form the reduction of induction to probability might take, there were certain inductive principles which had to be preserved, principles which had been articulated and established *prior* to the introduction of elaborate theories of probability into epistemology. These functioned effectively as *adequacy* conditions, spelling out certain minimal conditions which any probabilistic reconstruction of induction must preserve. Among these requirements[2] were:

(1) that the probability of any nonrefuted hypothesis must be between zero and one (which are thought to represent impossibility and certainty respectively);
(2) that the probability of an hypothesis must increase with the addition of confirming instances, but not in a linear fashion;
(3) that, as Whewell and Herschel had argued,[3] an hypothesis is rendered more probable by successfully explaining or predicting surprising phenomena than by the successful prediction of unsurprising phenomena;
(4) that the probabilities assigned to a scientific theory and a mere empirical generalization must, on comparable evidence, be different.

Any theory of probability which failed to satisfy these pre-formal conditions could not be considered as a legitimate 'explication' of the sense of 'probable' involved in the assertions of scientists about the probabilities of their theories. As we shall see, it was the failure of various theories of probabilistic induction to satisfy these conditons which resulted in the abandonment – at least for a time – of the program for reducing induction to probability theory.

One of the first significant attempts to recast induction in terms of probability theory was that of Augustus DeMorgan.[4] Briefly, he sought to apply the general division theorem and a form of Bayes' theorem to the problem of the probability of scientific hypotheses. DeMorgan maintained that these two principles of probability theory provided a criterion for choice between scientific hypotheses in terms of their posterior probabilities. George Boole took issue with DeMorgan in his *Laws of Thought*. Boole more or less concedes that Bayesian inference does succeed in reproducing many of our pre-formal intuitions about the probabilities of scientific hypotheses, and how those probabilities are affected by evidential considerations. However, Boole insists that DeMorgan's work is for nought because the use of Bayes' theorem involves two altogether *arbitrary* elements:

(1) the assignment of prior probabilities to hypotheses; and
(2) the determination of the value of $p(E/{-H})$.

Boole claims that these elements render Bayesian inference even more arbitrary and problematic than what he calls our "native intuitions" about the probabilities of scientific hypotheses. These weaknesses were sufficient to convince Boole that "the principles of the theory of probabilities [cannot] serve to guide us in the election of . . . [scientific] hypotheses".[5] DeMorgan's early program for reducing induction to mathematical probability attracted few adherents. Indeed, within the next decade, the probability theorists lose a succession of battles. Mill argues that probability theory is relevant only to the establishment of empirical generalizations, not to scientific laws, and that probability theory is therefore irrelevant to the logic of the mature sciences. Fries, pursuing another angle, argues in 1842 – and this became a common view – that the theory of probability deals only with the relative frequency of events, whereas what is required is a calculus of the probability of beliefs – which (Fries argued) is not reducible to relative frequencies. In drawing a distinction between what he calls 'mathematical probability' and 'philosophical probability' – a distinction which corresponds closely to Carnap's later distinction between probability_2 and probability_1 – Fries' argument that philosophical probability is inconsistent with mathematical probability is repeated by many critics of probability, including Apelt (1854), Cournot (1843), and Drobisch (1851).

By the early 1870s and perhaps not unreasonably, most inductive logicians had apparently abandoned the attempt to find a solution for the crisis of induction by reducing induction to probability theory. It was in this prevailing climate of opinion that Stanley Jevons challenged the prevailing orthodoxy by claiming, in his *Principles of Science* (1874), that "it is impossible to expound the methods of induction in a sound manner, without resting them upon the theory of probability".[6] Not by nature a slogan-monger, Jevons devotes much of his 800-page work to establishing that dependence. I want to look at Jevons' argument in some detail, as well as those of his critics, because the ensuing debate did much to shape the form in which the issue was raised afresh by Keynes, Reichenbach, and Popper in the second and third decades of the 20th century.

The chain of arguments whereby Jevons arrives at the conclusion that induction rests on probability is a curious one. He begins from the now commonplace but then important premise that the laws of science are merely probable, not certain. To this premise, he adds a theory of formal logic which distinguishes four forms of inference: deductive and inductive, probabilistic and nonprobabilistic. Probabilistic inferences are those which have statements of probability as premises or conclusions. Thus, probabilistic inferences may be either deductive or inductive and so may nonprobabilistic inferences assume either a deductive or an inductive form.

Having adopted this rudimentary taxonomy, Jevons proceeds to argue that induction "is simply an *inverse* employment of deduction".[7] A legitimate inductive inference is one which would produce a valid deductive inference if the conclusion and major premise were transposed. He then turns his attention to the two forms of probabilistic inference, the deductive and the inductive forms. The conclusions of deductive probabilistic inferences are certain, and their logic is that of ordinary probability theory. However, the conclusions of inductive probabilistic inferences are not certain. Since scientific inference is inductive and its conclusions are merely probable, and since induction is simply the inverse of deduction, Jevons concludes that scientific inference is the inverse of a probabilistic deduction. Jevons' problem then becomes that of determining the logical form of an inversion of a probabilistic deduction. But for this he believes he has a ready-made solution in the form of the theory of inverse probabilities as developed by Laplace.

Diagrammatically, one could represent Jevons' approach to inference this way:

	probabilistic	nonprobabilistic
Deduction	classical probability	syllogistic logic
Inductive	inverse probability	?

(By a confusion on Jevons' part, he manages to ignore altogether what is, for us as well as for Jevons' contemporaries, the most important species of scientific inference, viz., nonprobabilistic inductive arguments. I shall return to this omission later on.)

For now, let us look at how Jevons attempts to reconstruct inductive inference in terms of inverse probabilities. There are two principles of inverse probability which Jevons utilizes. One concerns the probability of causes. As Jevons puts it (in what is actually a paraphrase of Laplace):

> if an event can be produced by any one of a certain number of causes, all equally probable *a priori*, the probabilities . . . of these causes . . . are proportional to the probabilities of the event as derived from these causes.[8]

Jevons points out that this amounts to the assertion that if the possible causes of an event are equally probable *a priori*, that cause is most probable *a posteriori* which assigns the highest probability to the event. The second principle of inverse probability which Jevons utilizes is the famous set of

Laplacean rules of succession relating the probabilities of events to observed relative frequencies of their past occurrence.

With these tools in hand, Jevons attempts his promised reconstruction of the logic of scientific inference. And quickly, problems, even inconsistencies, appear, which Jevons only half recognizes. He sees, perhaps under prodding from DeMorgan, that the rules of succession entail that the probabilities of universal theories and hypotheses are vanishingly small, if they have any determinate value at all. Like Carnap 80 years later, Jevons seems at first willing to live with this situation. As he puts it, it is not unreasonable to think that "inferences pushed far beyond their data soon lose any considerable probability".[9] However, he has a low tolerance level for such a paradox when it is applied to universal scientific laws, for on the same page, he equivocates by insisting that, somehow, a sufficiently prolonged but finite experience will enable "the probability of the inductive hypothesis to approximate closely to certainty",[10] a view completely inconsistent with the rules of succession. As for Jevons' first principle, that concerned with the probability of causes, he *is* able to make a convincing case that it is sometimes utilized in scientific situations. But that principle suffers from other serious weaknesses. Never mind that it requires us to know the *a priori* probabilities of causal hypotheses. Even ignoring that, its irremediable weakness is that it is only appropriate for choosing between probabilistic hypotheses, i.e., hypotheses which are themselves probability statements. It offers no rank ordering for hypotheses which are equiprobable *a priori* and which assign a probability of one to the observed phenomena, for they are all equally probable on the given evidence. If an hypothesis *entails* the known evidence, as usually happens in cases of hypothetico-deductive inference, then we are forced by Jevons' causal principle to assign a probability of one to the hypothesis. Moreover, Jevons' rule makes it impossible to assess the changing probability of a single hypothesis as further supporting evidence emerges, a circumstance which both Peirce and Venn were quick to point out. For Jevons, a universal nonprobabilistic hypothesis is rendered as probable by a single confirming instance as it is by a hundred such instances. This point arises because of an earlier confusion on Jevons' part. Recall that when he was speaking of the four modes of inferences, he assumed that all inductive inference was probabilistic, i.e., that the conclusion of every inductive inference is a statement in the language of probabilities. This mistake made it possible for him to ignore that important class of inductive, nonprobabilistic inferences which does not readily fit into Jevons' version of the principle of inverse probabilities.

Indeed, if one looks closely at Jevons' examples, there are strong grounds for doubting whether he was seriously attempting to develop anything like a theory of confirmation. Without exception, Jevons' examples concern themselves, not with determining the probability of any specific universal hypothesis, but rather with determining the probability that a given set of phenomena are the result of some regular, causal agent. His concern, like Laplace's, was with problems like: What is the probability that the planets would move in the same direction by chance? What is the probability that the stellar distribution in our galaxy is random? What is the probability that the spectral lines from the sun would coincide with the spectral lines of hydrogren if the sun were not composed of hydrogen? In every case which Jevons analyzes in any detail, his concern is either with the probability of a single event or with the probability that a group of data exhibit some degree of regularity. In these examples, he never addresses himself to the problem of determining the probability of a particular universal hypothesis on the available evidence.

Jevons' *Principles* was scarcely off the presses before his purported reduction of induction to probability theory came under attack from several quarters. Chief among his critics was John Venn who devoted a chapter of his *Logic of Chance* (1888) to the connection between probability and induction. Referring contemptuously to that "curious doctrine adopted by Jevons, that the principles of induction rest entirely upon the theory of Probability",[11] Venn quarrels with Jevons' analysis on several counts. Specifically, Venn claims that:

(1) Every application of probability theory to concrete situations presupposes the validity of inductive inference, not vice versa;[12]
(2) Mathematical and philosophical probability are different, for the former deals with relative frequencies and the latter with rational expectations;
(3) The rule of succession has no use in inductive logic because "the strength of our conviction . . . will depend not merely on observed coincidences, but on far more complicated considerations";[13]
(4) At best, probability theory offers "the natural history of belief rather than its . . . justification";[14]
(5) Probability theory only offers rules for drawing deductive conclusions from probabilistic generalizations; it does not tell us how reliable those generalizations are, nor how well they are confirmed by the evidence;

(6) The inverse principle requires us arbitrarily to assign prior probabilities to hypotheses;

(7) The rules of succession entail that the probability of our most respected theories is zero, if they are interpreted – as Venn thinks they should be – as universal in scope;

(8) When we say, as inductive logicians, that a law is only probable, we are simply expressing "a vague want of confidence which cannot be referred to any statistical grounds for its justification, at least not in a quantitative way".[15]

As a result of these shortcomings, Venn – and evidently most of Jevons' other readers – are led to conclude that "it seems decidedly misleading to speak of [Inductive Logic] as resting or depending upon probability".[16]

Although Jevons never explicitly concedes defeat, he as much as admits that his reduction of inductive inference to probability theory fails. He does this by developing, alongside of his probabilistic theory of induction, a crude theory of *qualitative* confirmation for universal hypotheses, which has no direct linkage with the mathematical theory of probability. Indeed, in many ways, this is the most interesting aspect of the Jevonsian system, for it amounts to one of the earliest systematic statements of the hypothetico-deductive method and the theory of partial verifiability. In outlining that method, Jevons insists that all we can demand of our hypotheses is that they *agree* with the available evidence. Beyond such agreement, we cannot talk quantitatively about the degree of probability of any hypothesis, except to say about it just those four points I mentioned earlier as the intuitive, preformal characteristics of the probability of hypotheses. In the final analysis, Jevons, far from justifying those characteristics within the formal theory of probability, never even attempts to do so.

The paradox of Jevons' work, and at the same time the great strength of his analysis, is that his most sophisticated theory of scientific inference is one which he develops independently of, and with no reference to, that mathematical theory of probability which he originally held to be the touchstone of inductive inference. The Jevonsian dilemma is instructive, for it illustrates a phenomenon which has recurred more than once in the century since the publication of the *Principles of Science*. Like Jevons, Charles Sanders Peirce set out to justify inductive inference in terms of probability theory. Also like Jevons, though for very different reasons, Peirce's probabilistic reconstruction of the logic of science failed and Peirce returned to a nonprobabilistic approach to scientific inference.[17] More recently, the debates between Popper and Carnap have often turned on precisely the points which wee at stake between Jevons and his critics.

But it would take more space than I have here to explore some of these fascinating historical parallels. What I hope is clear is that 19th-century opposition to the attempt to reduce inductive inference to probabilities was not based entirely on blind prejudice. When Keynes asked fifty years ago how 19th-century authors could have dismissed the probabilistic approach to induction so hastily, he seems to have believed that Jevons' arguments for that approach were cogent and well reasoned. In taking such a view, Keynes – and certain more recent inductive logicians – have ignored just how weak Jevons' case was, and how little the Jevonsian probabilistic analysis contributed to an understanding of the nature of inductive inference and scientific logic.

How much significance to attach to this episode in evaluating the tradition of probabilistic induction is a matter beyond the scope of this chapter. But it is my distinct impression that when the more recent history of probabilistic induction is told, including the work of Keynes, Reichenbach, and Carnap, we will see a similar story of a tradition generating more problems than it resolves and often substituting formalistic elegance and a preoccupation with technical minutiae for a sophisticated analysis of the modes of scientific inference. To put it another way, I suspect that what I have been calling Jevons' dilemma might well turn out to be the dilemma of probabilistic induction generally.

NOTES

1 Had John Strong (who was writing the definitive study of these issues) not met such an untimely death, this chapter would have been suppressed – for it would have been completely superceded by his work.

2 Such requirements are formulated in a variety of ways by such writers as Whewell, Herschel, and Mill. It is important to realize that these requirements emerged *before* the application of probability theory.

3 See Chapter 10.

4 DeMorgan's treatment of this problem can be found in the article on 'Probabilities' in the *Encyclopedia Metropolitana*.

5 G. Boole, *An Investigation of the Laws of Thought* (London, 1854), p. 375.

6 S. Jevons, *The Principles of Science*, 2nd ed. (London, 1877), p. 197.

7 Ibid., p. viii.

8 Ibid., p. 243.

9 Ibid., p. 259.

10 Ibid.

11 J. Venn, *Logic of Chance*, 2nd ed. (London, 1888), p. 201.

[12] Ibid.
[13] Ibid., p. 359.
[14] Ibid., p. 201.
[15] Ibid., p. 211.
[16] Ibid.
[17] See Chapter 14.

CHAPTER 13

ERNST MACH'S OPPOSITION TO ATOMISM

INTRODUCTION

In those annals of history which record the noble espousal of lost causes, the name of Ernst Mach is often linked with the opposition to atomic and molecular theories, along with such figures as Ostwald, Stallo, and Duhem. And rightly so, for Mach's writings over a fifty-year period from 1866[1] until 1916 reveal an extreme suspicion of, bordering often on a hostility to, most of the atomistic theories of that period. Moreover, the running dispute between Mach and Boltzmann and later between Mach and Planck centered squarely on the efficacy of atomic and molecular approaches to the exploration of natural phenomena. But though the fact of Mach's opposition to atomic/molecular theories is well known and widely cited, Mach's specific argumentative strategies against such theories have been less fully explored and understood. Still less well documented is the relation of Mach's stand on atomism to his other work in philosophy, especially in the area of the logic and epistemology of science. Finally, the relation of Mach's critique of atomism to the views of his scientific and philosophical contemporaries is almost completely unexplored terrain.

It will be the aims of this chapter (1) to make a little clearer Mach's specific criticisms of atomism; (2) to suggest, albeit tentatively, the extent to which those criticisms do, and the extent to which they do not, link up with other portions of Mach's general views on the nature of scientific knowledge; and (3) to explore, in preliminary fashion, the intellectual traditions and affiliations to which Mach's critique of atomism reveals him to be allied. Perhaps the best place to begin is by surveying briefly the two major extant views about the origins and rationale for Mach's anti-atomism, in order to indicate why I find them unsatisfactory before I move on to state the positive case for the view of Mach I shall be defending. I shall label these positions the *sensationalist* explanation of Mach's anti-atomism and the *scientific* explanation of Mach's anti-atomism.

The Sensationalist Account

On this, the most common approach to the matter, it is argued that Mach's reservations about atomic and molecular theories derive from his sensationalist epistemology.[2] Mach, it is claimed, believed that the knowable world consisted solely of sensations (or, as he preferred to call them, 'elements') and their spatio-temporal contiguities and interconnections. Any reference to entities beyond sensation, to *Dinge an sich*, was illegitimate. All talk about external objects is just a shorthand way of speaking about our perceptions, actual or possible. Because atoms and molecules are, in principle, beyond the reach of our senses, because in short they are radically imperceptible, no theory that refers to atoms or molecules is meaningful.

There is much that is appealing in this approach to Mach. It provides immediately a neat linkage between his abstract philosophical concerns and his concrete physics, by thrusting Mach's *Contributions to the Analysis of Sensations* of 1886 to the center stage in any exegesis of Mach's thought. (It also, incidentally, makes it possible for recent philosophers to dismiss with a wave of the hand Mach's reservations about atomic doctrines by pointing out that those reservations rest on a discredited epistemological cliché: that the world consists entirely of sensations.)

But, for all its initial plausibility, this approach simply will not do as a *general* explanation of Mach's stand on atoms, nor will this interpretation bear the exegetical weight that it is forced to carry by those who see sensationalism as the philosophic cornerstone of Mach's approach to theory construction. Let me survey very briefly some of its weaknesses:

(1) Chronologically, it makes little sense because Mach raises serious doubts about the viability and efficacy of atomic theories long *before* he becomes a convinced sensationalist. Almost a quarter of a century before his *Analysis of Sensations*, and fully a decade before he develops the phenomenalistic epistemology which that work adumbrates, Mach was voicing the most serious reservations about atomic/molecular theories. Those who see Mach's sensationalism as the underpinning for his views on atomism must claim, in the absence of any substantial evidence, that he had adopted a sensationalist approach long before he published anything whatever on epistemological or generally philosophical questions.[3]

(2) Of greater concern is the fact that relatively few of the specific criticisms that Mach directs against atomic theory are couched in terms of a sensationalist theory of perception. As I shall show below, the bulk of Mach's reservations about atoms and atomic theories have nothing whatever to do with the irreducibility of atoms to sensations. His arguments come from different quarters, elsewhere in his philosophical system, and neither stand nor fall with the fate of his sensationalism.

(3) Most telling of all, however, is the fact that the sensationalist account of Mach's stand on atomism explains far *too much*, for it fails to indicate why it was atomic and molecular theories in particular that offended Mach's natural philosophical sensibilities. As Mach himself persistently points out, virtually *every* scientific theory postulates objects that transcend our sensory experience. To speak of oxygen, or heat, or a center of gravity, or a gravitational force, or even ordinary matter is to super-add or super-induce a conception of the mind onto our sensory experience. Even to refer to tables and chairs in the customary way as permanently enduring material objects is to involve oneself in the postulation of theoretical constructs that go well beyond the sensuously given.[4] Yet Mach does not recommend that we abandon or even seriously qualify our theories of chemical elements, or the sciences of statics or celestial mechanics. He has *no* general axe to grind against theorizing as such, nor against most of the scientific theories of his day, despite the fact that virtually all of them go well beyond what a sensationalist account of knowledge would legitimate. Clearly, if we are to understand what it was that specifically rendered atomic/molecular theories otiose in Mach's eyes, we must look elsewhere than to his sensationalism for the source of the anxiety.[5]

The Scientific Account

At the other extreme of Mach historiography are those scholars who are inclined to dismiss any attempt to find the source of Mach's stand on atomism in his philosophy and who point, instead, to certain important facets of Mach's *scientific* career that might explain his diffidence about theories of the micro-realm. Thus, in an important study of Mach's thought, Erwin Hiebert has suggested that Mach abandoned the atomism of his youth for two straightforwardly scientific reasons: (1) because Mach discovered certain empirical phenomena (especially concerning the spectra of elements) that seemed to defy atomistic interpretation and (2) because, after 1865, most of Mach's own scientific work on acoustics, on resonance and on psycho-physical problems was concerned with macro-level problems to which atomic/molecular theories seemed irrelevant.[6] Indeed, Hiebert goes so far as to claim that if Mach had been working in other domains of physics, "Mach very likely would have been more tolerant about atomistic conceptions".[7]

There is doubtless much in this approach to Mach, not least because it rightly stresses the developmental dimensions in understanding Mach's work. But it seems to me that, like the sensationalist account, it still fails to throw much light on some of Mach's basis reasons for opposing atomism. Let me here outline what seem to be its major deficiencies:

(1) If Hiebert is right that Mach's hostility toward atomic theories coincides with a shift in research concerns, then it seems difficult to understand why Mach, even when he is utilizing the atomic theory in his early work, voices serious doubts about its scientific credentials. As Hiebert himself has pointed out, Mach's early *Compendium der Physik fuer Mediciner* (1863) is as striking for the reservations it voices about atomic doctrines as it is conventional for its extensive use of such doctrines. Under such circumstances it seems implausible to explain Mach's aversion to atomism by invoking a shift from traditional theoretical physics to macro-level or psycho-physics, since Mach's reservations about atomic/molecular theories are voiced even in contexts where he is doing 'micro-physics'.

(2) If Mach's objections to atomism has been primarily experimental objections – as Hiebert repeatedly stresses they were – it is curious that on most occasions when Mach discusses atomic/molecular theories he fails to rehearse most of the known experimental weaknesses of such theories. In sharp contrast to most of the other opponents of atomism in the late 19th century – figures like Stallo, Helm, Brodie, Ostwald, and Mills – Mach rarely if ever goes through a detailed inventory of the empirical anomalies confronting the microphysical theories of his day. That is, I believe, because Mach does not care to show merely that atomic theories are false. Rather, he wants to show that they are inappropriate and dangerous as a species of theorizing and he needs more than a few refutations to establish that general thesis.

As a final caveat concerning Hiebert's view that Mach's stand in the atomic debates might well have been different if he had been working in other areas of physics or chemistry, we should remember that views on the propriety of atomic theories in the 19th century seem to have been as divided among those scientists working in fields where those theories were prominent as they were in fields far removed from atomic and molecular concern. Still more to the point, there seems to have been as much epistemic scepticism about the atomic theory from those very scientists who utilized it as from those who had nothing to do with it.[8] The fact that figures with scientific interests as diverse as those of Kelvin, Duhem, Kekulé, and Poincaré did, at various stages in their careers, voice grave doubts about atomic and molecular theorizing suggests that the sources of doubt had rather less to do with the problems on which a scientist was working than Hiebert allows.

But we now find ourselves in an interpretative vacuum. If it is neither abstract epistemology nor concrete physics that generates Mach's anxieties, where do they come from? It will be one of the central claims of this chapter that the answer to that question is that Mach's reservations arise not from his sensationalism, but rather from his methodology – that they come less from his theory of knowledge and more from his theory of science.

Put another way, my claim will be this: it is Mach's view about the *aims* and *methods* of scientific inquiry that provide the framework within which he develops most of his criticisms of atomic/molecular hypotheses. A more general corollary of this thesis is that if we wish to understand what was really at stake in the atomic debates that raged from 1860 to 1910, we must pay much more attention to *Wissenschaftstheorie*, particularly as it developed among Mach's scientific contemporaries, and rather less attention to *Erkenntnistheorie* as it developed among late 19th-century philosophers, including Mach himself.

Nor should this be surprising, particularly not in the case of Mach himself, who wrote approximately ten times as much on scientific methodology as he did on sensationalist epistemology. Indeed, almost all his major works are written either as explicit tracts on the methodology of science or else they contain lengthy sections on the appropriate methods and procedures of scientific inquiry. From *Die Geschichte und die Würzel des Sätzes von der Erhaltung der Arbeit* (1872), through *Die Mechanik* (1883), on to *Die Principien der Wärmelehre* (1896), *Populär-wissenschaftliche Vorlesungen* (1896), and finally to the magnificent but much neglected *Erkenntnis und Irrtum* (1905), Mach's continual preoccupation is with the nature of scientific inference, the relation of theory to experiment, the role of abstraction in theory construction, and other issues in the philosophy and methodology of science.

Moreover, there is abundant evidence that demonstrates that Mach's interest in methodology dates from a very early point in his career. As Wolfram Swoboda has shown in his study of the early development of Mach's thought, he was acutely concerned with questions of scientific methodology while still a young *Privatdozent* in Vienna. For instance, as early as 1860 (when he was 22) Mach became involved in a heated controversy with Josef Petzval about Doppler's study of the relation between frequency and pitch. In the course of that controversy he came to the conclusion that Pretzval's criticism of Doppler and of Mach hinged on a confused notion of the nature of analogical inference. Mach stated a resolve at that time to look more carefully into the logic of scientific inference. There is much evidence to suggest that he did just that, including a reference to Mill's *System of Logic* in a work of his in 1863 and a course of lectures he gave in Graz in 1864 on "The Methods of Scientific Research".[9] So far as his stand on atomism is concerned, I want to take Mach at his word when he warns us, "above all, there is *no* Machian philosophy, but at most a scientific methodology and a psychology of knowledge".[10]

MACH'S GENERAL THEORY OF METHOD

Before surveying Mach's specific arguments against atomism and their rationale, it would be helpful to recount some of the salient features of Mach's theory of science, particularily in the context of its ancestry through the 19th century. Even allowing for Mach's considerable eclecticism, his views on scientific knowledge and methodology can nonetheless be usefully classified as *postivistic*, provided that we take that term in its 19th- rather that its 20th-century signification.[11] For Mach, as for the positivists generally, the aims of science were descriptive and predictive. An ideal theory was one that, for the least labor, allowed one to represent as many known facts as possible and to anticipate or to predict correctly as many yet unknown states of the world as possible. Contrary to popular mythology, there was nothing in 19th-century positivism that was hostile to speculative theory construction.[12] All the major positivists from Comte to Mach, Poincaré, and Duhem enthusiastically accepted Kant's point about the active knower and were thoroughly contemptuous of that 18th-century brand of empiricism and inductivism which imagined that theories would somehow emerge mechanically from the data. Indeed, virtually all the 19th-century positivists, including both Comte and Mach, stressed that theories and hypotheses were a precondition for the coherent collection of experimental evidence.[13] In a perceptive passage in his *Science of Mechanics*, Mach formulates the point this way: after remarking that Galileo had certain "instinctive experiences" in his mind before he actually experimented on inclined planes, Mach stresses that

> for scientific purposes our mental representations of the facts of sensual experience must be submitted to *conceptual* formulation . . . this formulation is effected by isolating and emphasizing what is deemed of importance, by neglecting what is subsidiary, by *abstracting*, by idealizing . . . . Without some preconceived opinion the experiment is impossible . . . . For how and on what could we experiment if we did not previously have some suspicion of what we were about?[14]

But, as positivists kept stressing throughout the 19th century, even if theories are preconditions for the assimilation of data, theories are nonetheless *about* the data, about the facts, and it is those facts that provide the ultimate touchstone for choosing between theories. However, to put the point that way is to phrase it too mildly, for one of the most persistent themes in all of 19th- and early 20th-century positivism was the *conservative* nature of theorizing. Given that the aim of a theory is to correlate the facts, given that the process of correlation must take one beyond the facts, it should do so without postulating any other entities or processes than are *necessary* for that task of correlation. To develop a theory or hypothesis that, in order to codify and interrelate known phenomena, makes use of mechanisms or entities that are in principle beyond the reach of experimental analysis is to confuse science with pseudo-science. This requirement, that we should postulate no more than is necessary to explain the data, was called by Comte "the fundamental condition of hypothesis" and became, at the hands of Mach, the cornerstone of his doctrine of the economy of scientific thought.[15]

Closely linked to this ontological conservatism was a thesis of positivistic elimination – a kind of a Comtist razor. The thesis of elimination, as formulated by Mach and his predecessors, insisted that any theoretical entities that were not themselves subject to experimental analysis and control, that were not *verae causae*, were either to be eliminated from science altogether or else were to be treated as fictions, prophylactic devices that were themselves ontologically sterile. Thus, as Comte argued in the 1830s, the Huygensian construction of a wave front was purely a calculational device, a *façon de parler*, for reconstructing how light would move through a refracting interface. Secondary wavelets, so essential to that model, were purely ficticious. This conservatism, coupled with the doctrine of the elimination of fictions, made it natural for the positivists, including Mach, to distinguish between what we might call *purely observational theories* and *mixed theories* – the latter being those which referred, at least in part, to entities or properties that were in principle beyond the reach of experimentation. Positivists, let it be stressed, were not necessarily opposed to either type of theory; for there were certain circumstances in which even the most orthodox positivist was willing to endorse the use of hypotheses involving imperceptible, or purely theoretical, entities. (Comte, for instance, was sympathetic to the atomic theory of matter; Mach and Duhem to the undular theory of light.) But they objected strongly to efforts to put both types of theory on the same epistemic and methodological footing. To treat Fourier's or Carnot's theory of heat as being just like the kinetic theory of gases in all relevant respects was, they felt, to ignore the fact that the theories of Fourier and Carnot refer to nothing beyond what can be measured, whereas the kinetic heat theories of a Bernoulli, a Boerhaave, or a Herapath make continuous reference to entities beyond the reach of experimental analysis. Thus all the major positivists stress an important distinction within the class of acceptable theories – a distinction that has profound implications for our understanding of theoretical reference.

Within Mach's writings this distinction is most often formulated in terms of a distinction between universal theories that are *direct* descriptions (i.e., that contain only terms that give an abstract description of what is observable) and theories that are (or contain) *indirect* descriptions. Mach is willing to concede that either type of theory is scientific, although he himself has a decided preference for theories that involve only direct descriptions. His program moreover envisages the gradual replacement of indirectly descriptive theories by directly descriptive ones: it is, he asserts,

> not only advisable, but even necessary, with all due recognition of the helpfulness of theoretic ideas in research, yet gradually, as the new facts grow familiar, to substitute for indirect descriptions direct description, which contains nothing that is unessential and restricts itself absolutely to the abstract apprehension of facts.[16]

There is still another important positivist strain that is prominent in Mach's writings and that bears directly on his controversy with the atomists. From the time of Comte onward, the questions of the *classification of the sciences* had loomed large in positivistic and empiricist writings. Comte himself, Ampère, Mill, Cournot, and others had addressed themselves to the question of the logical and conceptual linkages between the various sciences. The standard positivist line, and one to which Mach usually subscribes (especially after 1865), is that the sciences do *not* generally stand in relations of logical deducibility to one another. Chemistry is *not* reducible to physics; biology does not follow from physics and chemistry; the social sciences are not derivable from the natural sciences.[17] This doctrine that each of the sciences has its own domain, its own concepts, and even its own methods is a very important one for Mach, and important at several levels. As psycho-physicist, he is keen to avoid the Fechnerian view that all psychical phenomena are intrinsically physical. Equally, as a physicist, Mach is adamant that the 'physical' is not exhausted by the science of mechanics. As we shall see below, Mach's view on the respective domains of the various sciences do much to condition his approach to the so-called atomic debates.

MACH'S CRITICISMS OF ATOMIC/MOLECULAR THEORIES

Much has been written in the last decade concerning Mach's view toward the so-called atomic-molecular hypothesis.[18] Some of this literature is flawed by a failure to distinguish a number of subtle but important differences in Mach's approach to this issue. Mach himself distinguishes between *physical* atomism and *chemical* atomism and has rather different observations to make about each one. More crucially, scholars have generally failed to distinguish between the rejection of an atomic hypothesis and the stating of reservations about an atomic hypothesis, too readily assuming that the latter entails the former. Again, there has been a tendency to assimilate Mach's views on the scientific weakness of the so-called *molecular* approach to his views on the metaphysical problems involved in an ontology of discrete entities.

But the ambiguity about the specific *objects* of Mach's attack has done less scholarly mischief than a failure to distinguish the various types of argumentative strategies that Mach deploys against atomic/molecular theories. Basically, Mach's arguments fall in four distinct classes:

(1) *The aim-theoretical argument*: objections or strengths here would be determined by showing the degree to which a theory was, or was not, conducive to, or compatible with, the accepted aims for scientific inquiry.
(2) *The interpretative argument*: a theory may be criticized for what its interpreters or partisans take it to have established or proved.
(3) *The programmatic argument*: a theory may be criticized or endorsed because it is part of a larger program of scientific research that is ill or well-conceived.
(4) *The inferential underdetermination argument*: the acceptance of a theory may be criticized if there are equally well-confirmed rivals to it.

Aim-Theoretic Objections

As is well known, Mach held deep-seated views on the aims of the scientific enterprise. In a phrase, Mach took the view that the aim of science was to describe and predict the course of nature as economically as possible. That meant, among other things, constructing theories that explained the known laws of nature and led to the discovery of new laws of nature by making the fewest possible existential commitments over and above those which the evidence warrants (or could conceivably warrant). In arguing that the aim of science is to find an economical description of nature, Mach is *not* suggesting that we must simply describe those natural regularities that we already know to be the case. It was of central importance for Mach that a theory must anticipate the future as much as recapitulate the past. Any theory worth its salt will lead the scientist to the discovery of new modes of interconnection between the appearances (i.e., to new laws). In order to achieve this aim, a theory may well go significantly beyond what we already know of the world and may make assertions about connections between phenomena that we have not yet explored. Far from viewing this process of going beyond the data as contrary to the aims of science, Mach stresses time and again the need for such theories.[19]

There are, in Mach's view, only two constraints on such theorizing. One is that such theories should lead us correctly to anticipate connections between observable data, i.e., they should predict some laws that later experiment may confirm. This I shall call the *weak constraint*. The second and *strong constraint* on such theories is that, if they are to be a relatively permanent part of our scientific ideology, we must be able to get some *direct* evidence for the existence of all the entities that such theories postulate.

Mach is not completely consistent concerning which of these constraints he will utilize in appraising micro-theories. In his most extreme moments he espouses the strong claim that a necessary condition for any satisfactory theory of physical processes is that *all* the connections postulated between entities in the theory must correspond to verifiable connections between physical objects or properties. As he formulates the thesis:

> In a complete theory, to all details of the phenomenon details of the hypothesis must correspond, and all rules for these hypothetical things must also be directly transferable to the phenomenon.[20]

Atomic/molecular theories, if assayed against this strong requirement of isomorphism, turn out to be counterfeit for, as Mach notes, "molecules (and atoms) are merely a valueless image",[21] in the sense that there are no experimental analogues for them or for their postulated modes of interaction.

But this extreme position, which would have literally legislated all atomic and molecular theories outside the domain of the sciences, is by no means Mach's most persistent or most characteristic stance. Much more often, Mach will allow that micro-theories are in principle acceptable, provided that they lead to the anticipation of new phenomenal or experimental laws; that is, so long as they exhibit a strong heuristic capacity.[22] Mach is well aware of the fact that some of the most predictively powerful theories in the history of science have involved the postulation of entities that were regarded as unobservable in principle. Far from unequivocally condemning the use of such theories, Mach has much to say on their behalf and frequently points out that the construction of such theories is often one of the best ways of discovering new facts.[23]

Thus, when Franklin sought to explain what was happening in the Leyden jar, he postulated an imperceptible, elastic electric fluid whose particles mutually repelled one another according to a $1/r$-repulsion law. Franklin's ideas about this (unobservable) electrical fluid led him to anticipate the fact that there are two modes of electrification (positive and negative), and that any change in one is accompanied by an equal and opposite change in the other. However, and Mach is insistent on this point, once we have established the functional dependence at the level of appearances, the theoretical model has served its purposes and can and should be discarded. Above all, Mach stresses, we must not confuse the tool with the job by pretending that the model does anything more than establish functional relationships between the data. As he put the point in 1890:

> The electric fluid is a thing of thought, a mental adjunct. [Such] implements of physical science [are] contrived for very special purposes. They are discarded, cast aside, when the interconnection ... has become familiar; for this last is the very gist of the affair. The implement is not of the same dignity, or reality ... and must not be place in the same category.[24]

Thus, for Mach, theoretical entities may play an important but intrinsically transitional role in natural science. Once they have suggested those empirical connections that are the warp and woof of scientific understanding, they can be discarded as so much unnecessary scaffolding. Insofar as the atomic theory helped scientists discover connections between the appearances (as Mach concedes it did with respect to the laws of definite and multiple proportions), it was of great heuristic use. But Mach considers that atomism long ago outlived its usefulness and is now simply redundant.

Like other positivists before him, Mach maintains that theoretical models are essentially temporary. In the course of time, either the models themselves become verifiable matters of fact or else they remain untestable and, having served to reveal whatever empirical connections they are capable of drawing attention to, are dropped in a favor of the lawlike, empirical connections themselves. Either way, they do not (or should not) remain as models for any significant length of time.

To the best of my knowledge, Mach *never* argues against the use of atomic and molecular theories, *provided* that such theories continue to lead us (as Dalton's did) to the discovery of new modes of connection between the data. That Mach acquiesces in the use of such theories does not mean that he is always happy with certain interpretations that certain atomists put on their theories (as we shall see below). But we cannot begin to understand Mach's reservations about atomic and molecular theories until we realize that he appreciates the heuristic potential that such theories have sometimes had and that he does not view the utilization of such theories as necessarily incompatible with the aims of science (so long as they anticipate new discoveries).[25]

Interpretative Objections

If Mach is relatively liberal and undogmatic about the *use* of atomic theories, he is substantially more adamant about the interpretation of such theories, particularly about interpretations that endow such theories with an ontological significance above and beyond the data that they correlate. Like Comte before him, Mach is concerned about the ontology of theoretical terms and concepts. In brief, Mach's position is that theories only make existential claims about those entities and properties which can be experimentally determined or measured. He develops this point at length in an important essay of 1890:

> A perfect physics could strive to accomplish nothing more than to make us familiar beforehand with whatever it were possible for us to come across (experimentally); that is, we should have knowledge of the interrelation of ABC . . . . A motion is either perceptible by the senses, as the displacing of a chair in a room or the vibration of a string, or it is only supplied, added (hypothetical), like the oscillation of the aether, the motion of atoms, and molecules, and so forth. In the first instance the motion is composed of ABC . . . , it is itself merely a certain relation between ABC . . . In the second instance the hypothetical motion, under especially favorable circumstances, can become perceptible by the senses. In which case the first instance recurs. But as long as this is not the case, or in circumstances in which this *can never happen* (the case of atoms and molecules), we have to do with a noumenon, that is, a mere mental auxiliary, an artificial expedient, the purpose of which is solely to indicate, to represent, after the fashion of a model, the connexion between ABC . . . .[26]

I take Mach to be arguing in this important passage that so long as it seems in principle impossible to get direct experimental evidence for the entities and modes of interaction to which a theory ostensibly refers, then we should limit the existential scope of that theory solely to its (in principle) observable entailments and view all other terms and concepts therein as purely fictional, as convenient algorithms for correlating data, which *mean* and *assert* nothing more than the observational connections which they entail.

It is crucial to distinguish this line of argument from Mach's views on the *use* of atomic/molecular theories. As we have seen, so long as such theories lead to the discovery of new empirically testable connections, Mach is more than willing to regard them as scientifically useful, regardless of whether we can 'observe' all the entities that the theories postulate. If, however, we cannot see any way to get direct empirical evidence for the existence of some of the entities to which the theories refer, then we should regard those entities as convenient fictions, not as laying bare a true ontology of nature. Atoms and molecules, in Mach's day, clearly fell into this category. He stresses this point at some length in a lecture in 1882 where he argues for the purely algorithmical character of atoms and molecules.[27]

It is important to stress again that Mach's anxieties are interpretative ones about ontology, not pragmatic worries about utilization. So long as atomic/molecular theories continue to produce discoveries of new macroscopic connections between observables, Mach would be the first to concede their value. It is only when scientists assume that atoms and molecules exist, even in the absence of any "direct" evidence for their existence, that Mach feels the interpretative machinery of science has gone astray.

Programmatic Objections

In Mach's view, specific attempts to formulate atomistic or molecular theories were a part of a more general reductionist program for the natural sciences. Indeed, Mach thought that much of the appeal of such doctrines to many theorists was precisely that they were inexorably linked with a program for reducing all of science to mechanics and all of the so-called secondary qualities to primary ones.

At this level of argument, Mach is not attacking atomic/molecular theories because they postulate unobservable entities but rather because such theories are inextricably bound up with what Mach regards as an outmoded and defunct research program for reducing all of nature to mechanics. His arguments here are fairly subtle, both historically and philosophically, so we must attend to them carefully.

As Mach points out, atomism has, at least since the middle of the 17th century, been tied up with two related *reductionist* doctrines. On the one hand, atoms have often been seen as the vehicle for effecting the reduction of all natural processes to the science of mechanics. Wherever mechanical philosophers observed any phenomenon that could not be directly reduced to the equations of motion (say, heat or chemical change), they immediately assumed that such phenomena – despite their apparently non-mechanical behavior at the macroscopic level – were the result of the motion of unseen atoms behaving as mechanical systems in miniature. Atoms thus functioned as *ersatz* equations of motion in the absence of the real thing. On the other hand, to look at the other tradition to which atomism was wedded, it has traditionally been associated with the primary-secondary qualities distinction and with the effort to reduce secondary qualities to primary ones, atoms again serving as the vehicle whereby the reduction was to be effected. As a historian of science, Mach saw clearly that atomism was the not very thin end of the wedge whose other apices asserted the primary of mechanics and the universality of the primary qualities. As he saw the science of his own time, atomism was being used to mechanize chemistry and gas theory and to deny the reality of that world of sensible qualities which Mach took to be fundamental.

Not surprisingly, therefore, many of Mach's arguments against atomic/molecular theories are directed against their mechanistic and reductive dimensions. As he argues at length in the *Science of Mechanics*, "the view that makes mechanics the basis of the remaining branches of physics, and explains all physical phenomena by mechanical ideas, is in our judgment a prejudice."[28] Mach urges his critics not to be confused by the historical accident that mechanics emerged first among the empirical sciences. Equally, he urges them to attach no more objective significance to "our experiences concerning relations of time and space than to our experiences of colors, sounds, temperatures and so forth".[29]

To seek to reduce everything to the behavior of mechanical atoms is prematurely to commit oneself to the general thesis that all change is mechanical and that all qualities are exclusively mechanical ones. Apart from the fat that such a thesis is unproven (and presumably unprovable), it suffers from other defects. As Mach points out in the *Science of Mechanics*, if it is part of the aim of science to explain the unfamiliar in terms of the more familiar, then it scarcely shows good sense always to seek to reduce the so-called secondary qualities of macroscopic bodies to conjectured primary qualities of hypothetical imperceptible bodies. Heat, light, work, pressure, and the like are palpable qualities of sensory bodies (and systems of bodies); and rather than explain them away by reducing them to atoms shorn of all such properties, Mach maintains that we should accept them as ultimately given and relate the behavior of such secondary qualities to others equally familiar in experience. Mach felt particularly strongly about the limits of mechanistic/atomistic science when it came to the biological and psychological sciences. The attempt by scientists such as Fechner to explain mental phenomena in atomic/molecular terms seemed particularly outrageous, even to involve what Mach calls a "flagrant absurdity".[30] A fuller understanding of the relations between physical and biological systems would, Mach believed, make one much less sanguine about the viability of reducing everything to a congeries of mechanical atoms. In his words:

an overestimation of physics, in contrast to physiology . . . a mistaken conception of the true relations of the two sciences, is displayed in the inquiry whether it is possible to *explain* feelings by the motion of atoms.[31]

What, in Mach's view, had made the mechanistic research program so attractive was the "misconception" that mechanical processes are more comprehensible, more intelligible, clearer than non-mechanical ones. Mach has two arguments against such Cartesian wish-fulfillment. For one thing, he maintains that most partisans of mechanistic reduction confuse intelligibility with mere familiarity. Because the science of mechanics has been around a good deal longer than (say) electrical theory or thermodynamics, people are more accustomed to utilizing it and to thinking of the world in terms of pulleys, levers, and inclined planes. But Mach argues that we no more 'understand' kinematic or dynamic processes than we 'understand' what is going on when a body cools or when we have a pain. All these processes are conceptually *unintelligible*. To seek to reduce everything to mechanics is merely to substitute what Mach calls a more "common unintelligibility" for a less common one. Mach underscores this point by stressing, in a more general vein, that the fundamental principles of any science, its "unexplained explainers", are always conceptually opaque, for they are the things posited as being beyond further analysis or comprehension. So, if we cannot render any of our basic conceptions intelligible, we must at least strive to make them factual by basing them on warranted abstractions from what we do know about the world. Atomic/molecular theories fail on this score, for they postulate entities that are neither intelligible nor experimentally verifiable. In his classic study on the *Conservation of Energy*, he puts it this way:

The ultimate unintelligibilities on which science is founded must be facts, or, if they are hypotheses, must be capable of becoming facts. If the hypotheses are so chosen that their subject can never appeal to the senses, . . . and also can never be tested (as is the case with the mechanical molecular theory), the investigator has done more than science, whose aim is facts, requires of him – and this work of supererogation is an evil.[32]

Clearly, if Mach is any sample to go by (and there is much to suggest that he is very typical in this regard), one cannot begin to come to terms with the so-called atomic debates in the 19th century unless one realizes that those debates were as much about the viability of the mechanical philosophy as they were about the existence of atoms.

Objections from 'Underdetermination'

Mach has two objections he often voices against the inferential moves that partisans of atomic and molecular theories frequently make. On the one hand, Mach points out, they often assume that the fact that their theories work is presumptive evidence for the truth of such theories. Mach observes that such a fallacious affirmation of the consequent is generally egregious, but particularly so in the domains of physics and chemistry, where we already know of the existence of non-atomic and non-molecular theories that are as well-confirmed by the data as atomic/molecular ones.

As early as 1863, Mach was using this argument against the claims of the atomists for the unique empirical adequacy of their particular *Weltanschauung*. In his *Compendium der Physik für Mediciner* he characterized the atomic theory – which he utilized extensively in that work – as simply one mode among many for handling the data of physics and chemistry, all of which are observationally equivalent. He likens these different modes to transcription from a polar to a rectangular system of coordinates. There is, Mach hints, no more reason to believe the atomic theory to be uniquely true and natural representation of the world than there is for believing a specification of a point's location by polar coordinates to be the uniquely appropriate mode of characterizing its position.

Mach elaborates on this point at some length in his *Erkenntnis und Irrtum*. He there argues that a scientific theory must satisfy two conditions: it must explain the facts with precision, and it must be internally consistent. There are, in principle, many different theories or conceptions of the material world that will satisfy both conditions. Under such circumstances, preference for one mode of conceiving the world over another is entirely arbitrary. More to the point, he stresses that the availability of a potentially large set of empirically adequate and logically consistent conceptual systems rules out of court any easy slide from the empirical well-foundedness of a theory to its truth:

Different concepts can express the facts in the observational domain with the *same* exactness. The facts must be distinguished from the intellectual images whose origins they have conditioned. The latter – the concepts – must be consistent with observation and moreover they must be logically consistent with one another. These two requirements are satisfiable in *numerous* ways.[33]

It should be clear after this brief exegetical survey that the bulk of Mach's reservations about atomic/molecular theories arose – neither from his sensationalism nor from his physics – but rather from his conception of the role and significance of theory in the natural sciences. Equally, it should be clear that Mach's stand on the atomic theories of his day was complex, characterized by a continuous spectrum of attitudes, depending upon whether he was dealing with atomism as "a working hypothesis", a "useful heuristic", a "proven physical theory", or as an established "research program". If this is true, it follows that a proper understanding of the atomic debates in general and of the Mach–Boltzmann controversy in particular, must go behind and beneath the scientific and technical details of the dispute so as to unpack the very divergent views about the aim and structure of theory that those debates reveal and on which they depend. We must realize that here, as elsewhere, scientific controversies are firmly rooted in a wide range of logical, methodological, and philosophical differences of the first magnitude.

MACH AND 19TH-CENTURY PHILOSOPHY OF SCIENCE

Although the aim of the chapter thus far has been to locate Mach's specific attitude toward atomic and molecular theorizing within the context of his general philosophy of science, it is equally important to attempt to determine the relation between Mach's views on the nature of science and those of his scientific and philosophical predecessors and contemporaries. In part, this is important in order to realize just how much Mach's views were part of several mainstream traditions in 19th-century physics, chemistry, and methodology. But it is also important in order to understand why Mach's views received as sympathetic and widespread a hearing as they did. There seems much evidence to suggest that Mach's analysis of the methodological deficiencies of 19th-century atomistic theories struck a responsive chord in *most* of his contemporary readers, and that Mach's reticence about atomistic speculation, although couched in terms of his particular philosophy of science, was neither perverse nor atypical of the views of his most reflective and able contemporaries.

A thorough study of the historical background to, and reception of, Mach's analysis of microphysical theories would require a full chapter in itself. All I shall attempt here is to give a sampling of the views of some of Mach's predecessors and contemporaries in order to illustrate the extent to which Mach was far from being a voice in the wilderness. It is already well known that there were a number of prominent 19th-century figures who were less than enthusiastic about the atomic theory. Among these were Duhem, Comte, and Poincaré in France; Ostwald, Helm, and Avenarius in Germany; Rankine, Stallo, and Brodie in England and America. To lump Mach together with these figures is, however, probably more misleading than helpful. For one thing, such a grouping tends to disguise some very significant differences among the "anti-atomists". Helm, Ostwald, Rankine, and Duhem were all "energeticists" whose primary objections to atomic/molecular theories were quite different from those of Mach – an elementary point that some historians have ignored at their peril. But more importantly, the grouping of Mach with the so-called anti-atomists leads one to ignore the degree to which Mach's general methodological worries about micro-theories were very broadly shared in the scientific community, shared alike by atomists and anti-atomists. Indeed, one of the unnoted ironies of the historical situation is that most of the atomists (e.g., Boltzmann) *accepted* Mach's methodological stand, and attempted to show – not that his methodology was wrongheaded – but rather that atomic theories could be legitimated *within* the framework of a generally Machist, generally positivistic philosophy of science. As a result, those who – like Einstein[34] – see the subsequent acceptance of the atomic/molecular hypothesis as a repudiation of Mach's general positivism are being less than fair to the actual exigencies of the historical situation.

From the early years of the 19th century onward, there was a broad general consensus about the methodological problems of micro-theorizing that can be summarized in the following fashion: the atomic theory is nothing more than a possible (or, sometimes, plausible) hypothesis; whether we should retain that hypothesis depends on its fecundity in leading to the discovery of new empirical relations; even if it is retained, we should be most reticent about asserting the actual existence *in rerum natura* of atoms or any other non-verifiable entities.

Thus, Berzelius in his *Lehrbuch der Chemie* of 1827 argued that atomic theories were a

> mere method of representation for the combining elements, through which we facilitate our understanding of the phenomena, but one does not thereby aim to explain the processes as they really occur in nature.[35]

A few years later, Auguste Comte articulated his theory of the "logical artifice" – specifically in connection with the atomic theory – which allows for the legitimate use of atomic and molecular theories, so long as one does not endow them with any objective reality. The organic chemist August Kekulé, though himself a frequent proponent of the atomistic hypothesis, conceded in 1867 that "I have no hesitation in saying that from a philosophical point of view, I do not believe in the actual existence of atoms."[36] Two years later, the chemist A. W. Williamson, also a proponent of the atomic theory, observed "that chemists of high authority refer publicly to the atomic theory as something which they would be glad to dispense with, and which they are rather ashamed of using".[37] The chemist E. J. Mills, in a passage that shows striking echoes of Mach, observes in 1871 that

> a phenomenon is explained when it is shown to be a part or instance of one or more known and more general phenomena. Isomerism is not, therefore, explained by assertions about indivisibles, which have neither been themselves discovered nor shown to have any analogy in the facts or course of nature.[38]

Mills continues:

> It would be a matter of the highest importance, one would imagine, especially on the part of experimental advocates [of atomism], to adduce, or at any rate to endeavor to adduce, an atom itself as the best proof of its own existence. Not only has this never been done, but no attempts have been made to do it; and it is probable that the most enthusiastic atomist would be the first to smile at such an effort, or ridicule the supposed discovery.[39]

> The atomic theory has no experimental basis, is untrue of nature generally, and consists in the main of a materialistic fallacy.[40]

Opinion among physicists was often even more scathing. In 1844, Michael Faraday pointed out that

> the word *atom*, which can never be used without involving much that is purely hypothetical, is often intended to be used to express a simple fact; but good as the intention is, I have not yet found a mind that did habitually separate it from its accompanying temptations [i.e., the temptation to think of atoms as real].[41]

Another English natural philosopher, Colin Wright, put the argument against atomism differently. As he saw it, the atomic theory had been

> a mechanical conception suited, doubtlessly, to an age when an accurate knowledge of facts was only beginning to exist . . . it is unnecessary to express any facts, and incompetent (without *much* patching and blotching) to explain many generalizations . . . it is undesirable that the ideas and language of this hypothesis should occupy the prominent and fundamental part in chemical philosophy now attributed to them.[42]

Several partisans in the atomic debates, on both sides of the fence, noted that most of the critics of atomism were positivists and that many of their specific criticisms of the atomic theory had already been adumbrated by Comte and his philosophical and scientific disciples.[43]

Thus we can see that, even among the most enthusiastic partisans of the atomic theory, there was a general acceptance that the methodological criteria by which Mach sought to evaluate such theories were sound and reasonable. Thus Ludwig Boltzmann and Max Planck both agreed with Mach that an economical representation of the facts is the central aim of science. Boltzmann even concedes Mach's point that atomic theory has often been "a retarding influence and in some cases has served [as] useless ballast".[44] Another prominent atomist, Adolphe Wurtz, is even more explicit about accepting Mach's yardsticks for theory evaluation. In his *La Théorie atomique*, he stresses that it is important not to confuse "facts and hypotheses". He goes on:

> We may retain the hypothesis [of atoms] as long as it permits us to interpret the facts faithfully; grouping them, relating them to each other and predicting new things, as long as, in a word, it will show itself fecund.[45]

Wurtz then goes on to show how, in his view, the chemical atomic theory has achieved just this. The important point here is that even among Mach's scientific opponents, there is still widespread acceptance of something very like the criteria for theory appraisal that Mach was espousing.

What such passages as these would suggest is that Mach's affinities with his contemporaries were considerably greater than the usual image of Mach as an eccentric crank would allow. In arguing for a fictional interpretation of atoms and molecules, in opposing the mechanistic program inherent in 19th-century atomism, in insisting that science is fundamentally descriptive, in demanding that micro-theories must be fertile at the level of observation and measurement, Mach was voicing not just his own but the anxieties of an entire generation of physicists and philosophers who were acutely concerned with the methodological credentials of some of the most widely utilized theories of the day.

NOTES

[1] Although Mach's scientific career began in the late 1850s, I am dating his opposition to the atomic theory from the mid-1860s. There is much evidence, unfortunately almost

all of it ambiguous, that suggests that until the early 1860s Mach was a partisan of atomic and molecular hypotheses. My own belief is that Mach held serious reservations about atomism from the beginning of his scientific career, but that point requires much more elaboration than I can give it here. Useful discussions of Mach's early work can be found in S. Brush, 'Mach and Atomism', *Synthese* 18 (1968), 192–215. Virtually all the claims of this particular paper concern Mach's views from about 1863 onward.

[2] For examples of this approach to Mach, see J. Blackmore, *Ernst Mach* (Berkeley, 1972), pp. 321 ff., and F. Seaman, 'Mach's Rejection of Atomism', *J. H. I.* 29 (1968), 389–93, both of whom see Mach's 'phenomenalism' as the source of, and the motivation for, his rejection of atomism.

[3] It should be pointed out that Mach had written a treatise on certain problems in psychophysics and the problem or perception in the early 1860s. However, given that Mach was very much under the influence of Fechner at this time, and given that Mach was to repudiate Fechner's views in the *Analysis of Sensations*, it is most unlikely that Mach in the 1860s adhered to anything like the sensationalism of that later work. (Unfortunately, the manuscript of his early treatise on psychophysics does not seem to be extant.)

[4] Mach stresses this point often himself. See, for instance, E. Mach, *Beiträge zur Analyse der Empfindungen* (Jena, 1886); Eng. trans., *The Analysis of Sensations*, (Chicago, 1914), p. 311. Again in E. Mach, *The Science of Mechanics*, 6th ed. (LaSalle, Ill., 1960), Mach argues that although atoms "cannot be perceived by the senses", that alone does not differentiate them from other objects since "all substances . . . are things of thought" (p. 589).

For yet another variant on this theme, see Mach's 1892 article in *The Monist*, where he argues that notions as divergent as "the law of refraction, caloric, electricity, light-waves, molecules, atoms and energy all *and in the same way* must be regarded as mere helps or expedients to facilitate our view of things" (p. 202).

Even Mach's persistent opponent, Ludwig Boltzmann, points out that Mach is aware that a phenomenalistic epistemology cannot differentiate between atomistic conceptions and physical-thing conceptions, so far as their epistemic well-foundedness is concerned. See L. Boltzmann, *Populäre Schriften*, (Leipzig, 1905), p. 142.

[5] I wish to make clear that, in denying that Mach's sensationalist epistemology had much to do with his stand on the cogency of atomic/molecular theorizing, I am *not* asserting a general claim about the independence of Mach's philosophy of science from his sensationalism. There are numerous points of contact between the two which deserve careful exploration. Equally, I am not asserting that all of Mach's reasons for opposing atomic/molecular modes of explanation were independent of his sensationalism, for that claim, too, would be misleading. My thesis, rather, is that the bulk of Mach's *stated* reasons for opposing such theories are independent of his theory of perception and of epistemic 'elements'.

[6] See E. Hiebert, 'The Genesis of Mach's Early Views on Atomism', in R. Cohen and R. Seeger (eds.), *Ernst Mach: Physicist and Philosopher* (Dordrecht, 1970), pp. 79–106.

[7] Ibid., p. 95.

[8] See especially the section below dealing with Mach's specific criticisms of atomism.

[9] See *Akademische Behörden, Personalstand und Ordnung der öffentlichen Vorlesungen an der k. k. Karl-Franzens-Universität zu Gratz*, Graz, 1863–66. Swoboda's study is

entitled, 'The Thought and Works of the Young Ernst Mach and the Antecedents to his Philosophy', Ph.D., dissertation, University of Pittsburgh, 1973.

[10] E. Mach, *Erkenntnis und Irrtum* (Leipzig, 1905), p. vii. I have used the 1917 edition.

[11] Although Mach does not often explicitly identify himself as a 'positivist', his writings are strongly positivistic in tone and content, and he was regarded by many of his contemporaries as one of the leading exponents of positivism. Toward the end of his life, he did concede that he was a "positivist". See E. Mach, *Die Leitgedanken* (Leipzig, 1919), p. 15.

[12] Comte, for instance, often spoke of "l'introduction, strictement indispensable, des hypothèses en philosophie naturelle" (A. Comte, *Cours de philosophie positive*, 6 vols. [Paris, 1830–42], 2: 434).

[13] Thus, Comte writes: "Car, si d'un côté, tout théorie positive doit nécessairment être fondée sur les observations, il est également sensible, d'un autre côté, que, pour se livrer à l'observation, notre esprit a besoin d'une théorie quelconque" (*Cours de philosophie positive*, 1: 8–9).

[14] Mach, *The Science of Mechanics*, p. 161.

[15] For a more detailed discussion of Comte's views, see Chapter 9.

[16] E. Mach, *Popular Scientific Lectures*, 5th ed. (LaSalle, Ill., 1943), p. 248.

[17] Here again, it is important to stress the chronology. In Mach's very early scientific writings, he is himself a reductionist, arguing for the reduction of all of physics to "applied mechanics" (E. Mach, *Compendium der Physik für Mediciner* [Vienna, 1863], p. 55); for the reduction of physiology to "applied physics," ibid., p. 1; and elsewhere, for the reduction of chemistry and psychology to mechanics. This is a view, and a program, that Mach repudiated during the mid-1860s and that he argued against for the rest of his life. We do not yet have any satisfactory account of this important shift in Mach's thought.

[18] See especially J. Blackmore, *Ernst Mach*; J. Bradley, *Mach's Philosophy of Science* (London, 1971); S. Brush, 'Mach and Atomism'; G. Buchdahl, 'Sources of Scepticism in Atomic Theory', *B. J. P. S.* **10** (1960), 120–34; E. Hiebert, 'The Genesis of Mach's Early Views on Atomism', M. J. Nye, *Molecular Reality* (New York, 1972); and F. Seaman, 'Mach's Rejection of Atomism'.

[19] For reasons that have never been clear to me, most of Mach's recent commentators have assumed that his doctrine that "science is description" precluded him from recognizing that there is any predictive element in science or that science can go beyond the known data. Harold Jeffreys, for instance, writes that "Mach missed the point that to describe an observation that has not been made yet is not the same thing as to describe one that has been made; consequently he missed the whole problem of induction" (*Scientific Inference* [Cambridge, 1957], p. 15). It is, I suspect, Jeffreys and others like him, such as Braithwaite (R. Braithwaite, *Scientific explanation*, [Cambridge, 1953], p. 348), who miss Mach's point. In stressing the view that the aim of science is description, Mach is contrasting it not with prediction but rather with *explanation* (in the sense of identifying the underlying metaphysical causes of things). Mach stresses time and again the extent to which theories must anticipate new data, the degree to which every scientific hypothesis goes beyond a mere description of the known facts. One has only to glance at a work such as *Erkenntnis und Irrtum* to see the extent to which Mach did recognize several vital epistemic differences between descriptions of known facts and predictions of unknown ones and, correlatively, the extent to which he attempts to face up to the problem of induction in its various guises.

[20] E. Mach, *History and Root of the Principle of the Conservation of Energy*, trans. P. Jourdain, (Chicago, 1941), p. 57.
[21] Ibid.
[22] As he pointed out in *Wärmelehre*: "Der *heuristische* und *didaktische* Werth der Atomistik ... soll keineswegs in Abrede werden" (E. Mach, *Die Principien der Wärmelehre*, Leipzig, 1896], p. 430n).
[23] Speaking of Boltzmann's use of the atomic theory, Mach notes that "Der Forscher darf nicht nur, sondern soll alle Mittel verwenden, welche ihm helfen können" (ibid.).
[24] E. Mach, 'Some Questions of Psycho-physics', *Monist* 1 (1890), 393 ff., 396.
[25] Mach explicitly points out in *Wärmelehre* that "it should be emphasized that an hypothesis can have great heuristic value as a working hypothesis, and at the same time be of very dubious epistemological value" (*Die Principien der Wärmelehre*, p. 430 n).
[26] E. Mach, 'Some Questions of Psycho-physics'.
[27] The passage is probably worth quoting in full: "When a geometer wishes to understand the form of a curve, he first resolves it into small rectilinear elements. In doing this, however, he is fully aware that these elements are only provisional and arbitrary devices for comprehending in parts what he cannot comprehend as a whole. When the law of the curve is found he no longer thinks of the elements. Similarly, it would not become physical science to see in its self-created, changeable, economical tools, molecules and atoms, realities behind phenomena ... The atom must remain a tool for representing phenomena, like the functions of mathematics. Gradually, however, as the intellect, by contact with its subject matter, grows in discipline, physical science will give up its mosaic play with stones and will seek out the boundaries and forms of the bed in which the living stream of phenomena flows" (E. Mach, *Popular Scientific Lectures*, 5th ed. [LaSalle, Ill., 1943], pp. 206–7).
[28] Mach, *The Science of Mechanics*, p. 596.
[29] Ibid., p. 610.
[30] Ibid.
[31] Ibid.
[32] E. Mach, *History and Root of the Principle of the Conservation of Energy*, p. 5.
[33] Mach, *Erkenntnis und Irrtum*, p. 414.
[34] See, for instance, the discussion quoted in S. Suvorov, 'Einstein's Philosophical Views and Their Relation to His Physical Opinions', *Soviet Physics Uspekhi* 8 (1966), 578–609.
[35] Quoted in G. Buchdahl, 'Sources of Scepticism in Atomic Theory'.
[36] A. Kekulé, 'On Some Points of Chemical Philosophy', *Laboratory* 1 (1867), 304.
[37] A. Williamson, 'On the Atomic Theory', *Jour. Chem. Soc.* **22** (1869), 328.
[38] E. Mills, 'On Statistical and Dynamical Ideas on the Atomic Theory', *Phil. Mag.*, 8th ser., **42** (1871), 112–29.
[39] Ibid., p. 123.
[40] Ibid., p. 129.
[41] M. Faraday, 'Speculation Touching Electric Conduction and the Nature of Matter', *Phil. Mag.*, Ser. 3, **24** (1844): 136.
[42] C. Wright, *Chemical News* **24** (1874), 74–5.
[43] As Brock has pointed out, in W. Brock (ed.), *The Atomic Debates* (Leicester, 1967), pp. 145 ff., many of the most vocal members of the anti-atomist camp in

the 1850s and 1860s were followers of Comte, including Berthelot, Wyrouboff, and Naquet.

44 Boltzmann, *Populäre Schriften*, p. 155.

45 C. A. Wurtz, *La théorie atomique* (Paris, 1879), p. 2.

CHAPTER 14

PEIRCE AND THE TRIVIALIZATION OF THE SELF-CORRECTIVE THESIS*

If science lead us astray, more science will set us straight.[1]
– E. V. DAVIS (1914)

The aims of this chapter are two-fold: first and primarily, to identify and to summarize the development of an important but hitherto unnoticed tradition in 19th-century methodological thought, and secondly, to suggest that certain aspects of the history of this tradition give us a new perspective from which to assess certain strains in contemporary philosophy of science. In Part I below, I attempt to define this tradition, to document its existence, and to note some features of its evolution. In Part II, I briefly indicate the manner in which this history may shed new light on some recent trends in inductive logic.

I

As the title of the chapter suggests, the tradition that interests me is connected with the view of scientific inference as self-corrective, and the work of Charles Sanders Peirce looms large in the story.[2] It has been customary to see Peirce as the founder and first promulgator of the view that the methods of scientific inference are self-corrective.[3] This historical claim is simply incorrect. The doctrine that scientific methods are self-corrective, that science in its development is inexorably moving closer to the truth by a process of successive approximation, has a pedigree extending back at least a century before Peirce's time. And, in my view, Peirce's importance resides not in the creation of this doctrine but in his transformation of it in subtle but significant ways. As I shall argue below, Peirce is the crucial logical and historical link between 19th- and 20th-century discussions of self-correction and progress towards the truth. Moreover, he is responsible for effecting a major metamorphosis in the self-correcting doctrine as it had been understood by his predecessors. To get some sense of the magnitude of that mutation, we must go back to the middle of the 18th century to see how and why the idea of self-correcting modes of inference arose.

Beginning in the 1730s and 1740s, a number of philosophers and scientists began to claim that science, as a result of the methods it employs, is a self-corrective enterprise. (Hereafter I shall refer to this view as the self-corrective thesis or simply SCT.)

Most early versions of SCT – like their more recent counterparts – were closely connected with a theory of scientific progress (SCT asserting, in effect, that science does "progress") and it is, therefore, not surprising that the Enlightenment view of intellectual progress first provided a leitmotif and rationale for SCT. That eclectic theory of knowledge, unique to the *philosophes*, which identified the growth of the mind with the moral improvement of mankind, was certainly related to the doctrine of self-correction. But it is important not to be too beguiled by facile historical plausibilities. That the Enlightenment theory of progress produced fertile ground for the growth of the self-corrective view is quite likely; but we must look beyond the ethos of the Enlightenment to find the initial stimulus for theories of self-correction. Specifically, we must look to certain tensions and problems latent in the history of methodology itself. For instance, it is crucial to realize that the self-corrective thesis was itself a weakened form of a still more sweeping thesis which had dominated metascientific thought from antiquity. According to this more general thesis, which we might call the *thesis of instant, certain truth* (TICT), science – in so far as it is genuine science – utilizes a method of investigation which infallibly produces true theories. Virtually every theorist of method in the 17th century (including Bacon, Descartes, Locke, and Newton) subscribed to TICT.[4] The proponents of TICT believed that science could dispense with conjectures and hypotheses since there was, ready at hand, an "engine of discovery" (as Hooke called it)[5] which could infallibly (and usually mechanically) produce true theories. The concept of progress, within the framework of TICT, was clear and unambiguous. Progress, on this view, could only consist in *the accumulation of new truths*. The replacement of one partial truth by another simply made no sense in this context. Growth, in so far as it occurred, was by accretion rather than by attrition and modification.

By the middle of the 18th century, however, many methodologists were convinced that TICT was untenable. Difficulties in articulating a coherent logic of discovery, along with sceptical arguments about the inability of empirical evidence to prove a theory conclusively, conspired to chasten the scientist's confidence in the undisputed truth of his mental creations and to make (merely probable) hypotheses repectable, for the first time since the euphoria of the Scientific Revolution had made them unfashionable.[6]

There were two major arguments which seemed to undermine TICT: one was directed against the method of "proof *a posteriori*" (as Descartes had called it); the other, against eliminative induction. The main argument of the first kind was an application of the so-called "fallacy of affirming the consequent" to scientific inference. As surprising as it might seem, several methodologists and scientists in the 17th and 18th centuries had argued that the ability of a theory to predict successfully an experimental result was *prima facie* evidence that the theory was a proven truth. Cartesians (e.g., Jacques Rohault) and Newtonians (e.g., Bryan Robinson) alike often slipped into this sloppy mode of reasoning. By 1750, however, the inconclusive character of this form of inference had been pointed out by Leibniz, Condillac, and David Hartley, among others.

Similarly, the method of proof by eliminative induction (associated with Bacon and Hooke) had been discredited by the arguments of Condillac, Newton, and LeSage against the possibility of exhaustively enumerating all the conceivable hypotheses which might be invoked to explain a class of events. These three all asserted that (in light of the impossibility of knowing that we have thought of all the appropriate hypotheses which might explain facts in a given domain) we can never be sure that the hypotheses which have survived systematic attempts at refutation are true.

(The third major candidate for a model of scientific inference, *enumerative* induction, had long since been discredited; in antiquity by Aristotle and Sextus Empiricus, and in early modern times by Bacon, Newton, and Hume, among others.)

Since none of the known modes of "empirical inference" were valid, methodologists of science in the late 18th century were no longer able to speak, with a clear conscience, about the certainty and truth of scientific theories. (The notable exceptions to this generalization are the "a priorists" e.g., Lambert, Wolff, and Kant; but theirs was a minority viewpoint.)

Prepared to concede that the theories of the day might eventually be refuted, convinced moreover that TICT was too ambitious, several late 18th-century methodologists produced a compromise. If, they reasoned, there is no instant, immediate truth, we can at least hope to reach truth *in the long run*. Even if the scientist's methods do not guarantee that he can get the truth on the first attempt, perhaps he can at least hope to get ever closer to it. Even if the methods of science are not foolproof, perhaps they are capable of correcting any errors the scientist may fall prey to. Thus was born SCT. In some ways, it was a face-saving ploy, for it permitted the scientist to imagine that his ultimate goal was, as TICT had suggested, the Truth; although the scientist now had to be satisfied with the quest for ever-closer approximations rather than the truth itself.[7]

At the same time that SCT was emerging (and this was no coincidence) some methodologists were moving away from a Baconian inductive model of scientific inference towards something like a model of conjectures and refutations.[8] Science was seen, not as a discipline where theories were somehow extracted or deduced from experiment, but rather as one where theories were formulated, tested, rejected, and replaced by other theories. When SCT was stated within the context of such a model of scientific inquiry, it generally amounted to the following claims:

(1) Scientific method is such that, in the long run, its use will refute a false theory;

(2) Science possesses a method for finding an alternative T' which is closer to the truth than a refuted theory.[9]

On this view, which is as much an historiography as a philosophy of science, the temporal sequence of theories, in any genuinely scientific domain is a series of ever-closer approximations to the truth (provided of course, that science uses the method(s) which insure(s) self-correction). And there was a certain amount of intuitive plausibility to this picture. Even today, it is common to hear that Ptolemy's system was closer to the truth than Aristotle's system of concentric spheres; that Copernicus' helocentric system was 'more nearly true' than Ptolemy's; and that Kepler's elliptical system is a still closer approximation.

It is important to be clear about the set of problems which SCT was presumed to resolve. Like TICT before it, *SCT was designed to provide an epistemic solution to the problem of scientific knowledge*. That problem can be put in various forms: Why should we take science seriously as a cognitive pursuit? What justification is there for the methods which science employs? Why should we prefer science to quackery or pseudo-science? Whatever our views about SCT, we must at least concede that it was an attempt to resolve what are perhaps the central problems of the philosophy of science; namely, the justification of both the knowledge-claims and the methods of the natural sciences.

Adherents of SCT provided what was, in its time, a highly original approach to this perennial problem. For them, the justification of science as a cognitive, intellectual pursuit was sought – not in the certainty or even the truth of its conclusions – but in its progressive evolution towards the truth. As I shall show below, in the course of the later evolution of SCT, there was an increasing tendency to lose sight of this justificational problem in its full generality, a tendency to see the self-corrective thesis as the solution to rather different problems, of far less significance. But more of that below.

If the conditions I have spelled out indicate roughly what SCT amounts to, what was its rationale? What reason had Enlightenment philosophers to believe that science uses methods which satisfy conditions (1) and (2) above? The early proponents of SCT provided an answer, but not a very satisfactory one. Pursuing analogies between certain methods of mathematical inference and the methods of science, they claimed that just as the mathematician finds the roots of an equation by posing incorrect guesses, and then refines those guesses via mechanical tests, so the scientist can formulate an incorrect hypothesis and subsequently improve on it by comparing its results with observation, altering the hypothesis where necessary to bring it into closer agreement with fact. Clearly, the analogy is incomplete. After all, it does not prove that the methods of science are self-corrective to compare them with self-corrective mathematical techniques unless the analogies between the two cases are very strong in appropriate respects. Unfortunately, they are not (or, at least, they were not shown to be) strongly analogous. Although it is relatively easy to show that the method of hypothesis satisfies condition (1) above, there is no machinery for insuring that such a method satisfies condition (2) or even (2′).[10] Indeed, no methodological procedure was suggested in this period for replacing a refuted hypothesis by one which could be known (or reasonably presumed) to be closer to the truth. So impressed were these methodologists by the approximative techniques of mathematics that they did not worry about what (in our view) are the vast logical gaps between scientific testing and mathematical proof. This perhaps can be made clear by discussing a pair of representative early defenders of SCT. Among the first philosophers[11] to address themselves to this problem were David Hartley (1705–57) and Georges LeSage (1724–1803), who, although working independently, arrived at almost identical results. I shall consider them in turn.

In a chapter, "Of Propositions, and the Nature of Assent", in his *Observations on Man* (1749), Hartley analyzed the sorts of methods which the scientist has at his disposal. Hartley insisted that only in mathematics can one develop theories which can be rigorously demonstrated.[12] In science, however, we must be content with something less than certainty. However, taking his cue from the mathematicians, Hartley believed that the scientist can utilize certain methods which, if they do not yield the truth immediately, will gradually bring the scientist to a true theory in the long run. He proposed two different methods, both based on mathematical techniques, both of which are self-correcting, and both of which are, in the long run, supposed to lead the scientist to the truth:

(1) *The rule of false position.* This approximative technique, known as the *regula falsa* among Renaissance mathematicians, was characterized by Hartley as follows:

> Just as the arithmetician supposes a certain number to be that which is sought for; treats it as if it was that; and finding the deficiency or over-plus in the conclusion, rectifies the error of his first position by a proportional addition or subtraction, and thus solves the problem; so it is useful in all kinds of inquiries, to try all such suppositions as occur with any appearance of probability, to endeavour to deduce the real phenomena from them; and if they do not answer in some tolerable measure, to reject them at once; or if they do, to add, expunge, correct, and improve, till we have brought the hypothesis as near as we can to an agreement with nature. After this it must be left to be further corrected and improved, or entirely disproved . . . .[13]

Two centuries earlier, the mathematician Robert Recorde had, like Hartley, been impressed and amazed at the capacity of the rule of false position to generate truth from error, as this delightful piece of doggerel verse indicates:

> Gesse at this woorke as happe doth leade
> By chaunce to truthe you may procede
> And first woorke by the question,
> Although no truthe therein be don.
> Such falsehode is so good a grounde,
> That truthe by it will soone be founde.[14]

Hartley took Recorde's point one important step farther, however, by arguing that this sort of method works in natural philosophy as well as in algebra.

(2) *The method of approximating to the roots of an equation.* Like the rule of false position, this Newtonian technique was seen by Hartley as a means of generating a theory "which though not accurate, approaches, however, to the truth".[15] Here, the scientist begins by a guess at the root of the equation. From such a guess, applied to the equation, "a second position is deduced, which approaches nearer to the truth than the first; from the second, a third, etc."[16] Hartley insists that the use of such self-corrective methods "is indeed the way, in which all advances in science are carried on".[17]

There are, I believe, two important points to note about each of these methods. In the first place, both involve the inquirer in making posits (viz., hypotheses) which, if false, can be eventually falsified. Much more importantly, they both provide a method, having once refuted an hypothesis, for mechanically finding a replacement for it which is closer to the truth than the original hypothesis. These two characteristics together constitute the necessary and sufficient conditions for what I shall call a *strong self-correcting method* (or *SSCM*).[18] A method is an SSCM if and only if (a) it specifies a procedure for refuting a suitable hypothesis, and (b) it specifies a technique for replacing the refuted hypothesis by another which is closer to the truth than the refuted hypothesis.[19] Much of this paper will be concerned with post-Hartleyan accounts of SSCMs.

Unfortunately, Hartley himself did not indicate how we can apply such mathematical methods to the natural sciences. While it is easy enough to imagine that scientific hypotheses are refutable (neglecting Duhemian considerations), it is more difficult to guess what rule he had in mind for replacing a refuted scientific hypothesis by one which was more nearly true.[20] Hartley simply took it for granted that one can, in a more or less straightforward fashion, import these mathematical techniques into the logic of the natural sciences.

Hartley's contemporary, Georges LeSage, though drawing on slightly different mathematical analogies, made an argument very similar to Hartley's. LeSage compared the procedure of the scientist to that of a clerk solving a long-division exercise. At each stage in the division, we produce in the quotient a number which is more accurate than the number appearing as the quotient in the preceding stage. At each stage, we multiply the divisor by the assumed quotient and see if it corresponds to the dividend. If it does not, we know that there is an error in the quotient, and we have a mechanical process for correcting the error, i.e., for replacing the erroneous quotient by one which is closer to the true value.[21] Going beyond such fanciful examples, LeSage, like Hartley, suggested that there are other approximative techniques which the scientist can borrow from the mathematician, including "the extraction of roots, the search for the rational divisiors of an equation and several other arithmetical operations".[22] Beyond this, LeSage's views are, even to their ambiguity, sufficiently similar to Hartley's not to require separate consideration.

As I hinted before, the thesis that science is self-corrective and thereby progressive lends itself neatly to the 18th-century view of progress, for the sequence of theories of ever greater verisimilitude was the mirror image on the intellectual level of man's progressive perfection on the moral level. Joseph Priestley, who was in these matters a self-avowed disciple of Hartley, made explicit the link between the self-corrective character of science and his theory of scientific progress. He wrote:

Hypotheses, while they are considered merely as such, lead persons to try a variety of experiments, in order to ascertain them. These new facts serve to correct the hypothesis which gave occasion to them. The theory, thus corrected, serves to discover more new facts, which, as before, bring the theory still nearer to the truth. *In this progressive state, or method of approximation, things continue* . . . .[23]

Clearly, the weakness with all these programmatic statements is that they simply insist that scientific methods *are* self-corrective, without indicating precisely the manner in which they are so. Without a persuasive reason for believing that the methods of science are self-corrective, we have no rational grounds for speaking of scientific progress, a point which the logician and physiologist Jean Senebier was quick to emphasize: "Often we move imperceptibly away from the truth, and do so even whilst we believe that we are working towards it."[24] The case against the vagueness of SCT as developed by Hartley and LeSage was put convincingly by Pierre Prevost in 1805. He insisted that scientific procedures necessarily differ from such mathematical techniques as the rule of false position. He observed that we do not generally have the knowledge in science to be able to satisfy the conditions of the rule of false position, and that we therefore cannot expect much from that method in science. Prevost argued specifically against the self-correcting character of the method of hypothesis. All that method permits us to do, in his view, is verify or refute an hypothesis; it provides no machinery for replacing a refuted hypothesis with a better one:

Thus when Kepler, beginning with the circular hypothesis, tried out various eccentricities for the orbit of Mars, these false suppositions could (and indeed should) never have led him to a solution. When afterwards he recognized the weakness in the circular hypothesis, if he had tried other curves entirely by chance, he would have been using another method which could well have never brought him to his goal.[25]

I hope these few texts have made it reasonably clear that by the early years of the 19th century, the problem of justifying scientific knowledge (i.e., as infallible, indubitable truth) had been replaced – at least among some writers – by a program for justifying science by claiming that it pursues a method which will lead it ever closer to the truth. The extent to which this kind of approach quickly came to dominate methodological thought is illustrated by the fact that the philosophies of science of Herschel, Comte, and Whewell were all concerned overtly with the progress of science and its gradual approach to the truth.[26]

Among 19th-century scientists as well as methodologists, the view persisted of science as an enterprise moving inexorably closer to a final truth. Claude

Bernard, among others, conceived science in this approximative way. Thus, Ernest Renan wrote about Bernard:

> Truth was his religion: he never had any disillusionment or weakness, for not a moment did he doubt science . . . The results of modern science are not less valuable for being acquired by successive oscillations. These delicate approximations, this successive refining, which leads us to modes of understanding *ever closer to the truth* are [for Bernard] the very condition of the human mind.[27]

Similarly, that fervent Darwinian T. H. Huxley believed that "the historical progress of every science depends on the criticism of hypotheses – on the gradual stripping off, that is, of their untrue or superfluous parts . . . . "[28]

The key to the progressiveness of science was thought to reside in the fact that it utilized a method which was essentially self-corrective in character. Given time and sufficient experience, science could be perfected to any stage desired. In the middle years of the 19th century, especially with Comte and Whewell, the doctrine of progress through self-correction became, in many ways, the central concern of the philosophy of science. Science was seen as a growing, dynamic enterprise and, accordingly, philosophers of science were prone to stress such dynamic, growth-oriented parameters as increasing scope and generality, greater accuracy and systematicity and, above all, progress towards truth. However, throughout much of the 19th century, the self-corrective character of scientific method, while regularly invoked and persistently praised, remained as unestablished as it had been with LeSage and Hartley. Everyone assumed that science is self-corrective (and thereby progressive), but no one bothered to show that any of the methods actually being proposed by methodologists are, in fact, self-corrective methods.

The focus of the self-correcting thesis had always been on conceptual change, on the progressive succession of one theory by another. What self-correctionists had sometimes ignored was that sort of "progress" which comes from increasing the probability of theories (most often by successful confirmations), without any change in the theories themselves. This second type of progress, which we might call "progress by probabilification", received much attention in the 19th century. Herschel, Brown, Whewell, Jevons, and Apelt (to name only a few) discussed at length the methods by which we can gain confidence in our theories, without necessarily altering them. Partisans of progress through probabilification, tended to stress the continuity of scientific theory; for them, experiments with high confirming potential were emphasized rather than the falsifying experiments which the self-correctionists stressed. If the advice of the self-correctionists to experimental scientists was "Devise experiments which will indicate weaknesses in your theories," the corresponding advice from the probabilifications was "Devise experiments which, if their outcome is favorable, will do most to contribute to the likelihood of your theories." Impressed by Laplace's rule of succession and the application of probability theory to induction, the "probabilists" argued that every valid theory goes through all the degrees of certainty from extreme improbability to great likelihood. (Writers like Thomas Brown and John Herschel identified that transition as one from "hypothesis" to "theory" or "law".)

It would be wrong to give the impression that these two alternative theories of scientific progress, one by self-correction and the other by probabilification, were mutually exclusive. On the contrary, several of the best-known methodologists of the period (e.g., Whewell and Bernard) adopted both, arguing that "local" progress occurred by probabilification, while "cross-theoretical" progress was governed by a self-corrective method.[29] These two approaches did, however, represent different emphases, and were to give rise in the 20th century to two very different strains in philosophy of science (Carnap and Keynes being the descendants of the progress by probabilification school, and Popper and Reichenbach focussing primarily on progress by self-correction).

A third theory of scientific progress prominent in the 19th century was that endorsed by Mill and Bain. Mill adopts a theory of progress by elimination. An hypothesis is entertained, tested, refuted and replaced by another one. This perhaps seems but another version of the standard method of hypothesis. But it receives an interpretation by Mill very different from that of the self-correctionists. Mill does not believe we have any good ground for believing that a replaced hypothesis is any more true than its refuted predecessor. Indeed, it may be "more false". But the sequence of hypotheses is a progressive one, according to Mill, because the *last* remaining member of the series is true. Adhering to a principle of limited variety, Mill maintained that there was only a finite number of candidates for the status of a scientific law and the false contenders could be eliminated by a judicious use of the five canons of induction. Clearly, Mill's account of scientific progress differs substantially from that of both the self-correctionists and the probabilificationists.

All three of these theories of scientific progress found their followers in the second half of the 19th century. Nonetheless, the self-correctionists predominate, and it is late 19th-century developments in the self-corrective tradition which I want to examine now.

II

As we have seen, for more than a century after Hartley and LeSage, methodologists almost to a man (Mill being the most noteworthy exception) endorsed SCT and, ignoring the doubts voiced by Senebier and Prevost, assumed without much argument that the methods of science in general, and the method of hypothesis in particular, were genuinely self-corrective. The discussion of this question was given an entirely new slant, however, by the work of Charles Sanders Peirce, whose approach to this question I wish to discuss in some detail.

It is well known that Peirce was a persistent defender of SCT. Unlike most of his predecessors, however, Peirce (usually) realized that SCT was not self-evidently true, and felt that one of the tasks of the logician of science was to show how and why science is a self-corrective enterprise which, in its historical development, gradually but inexorably comes closer and closer to a true representation of natural phenomena.[30]

Peirce's most crucial claim in this regard in his insistence that *all scientific inquiry is self-corrective in nature*. "This marvelous self-correcting property of Reason", he wrote, "belongs to every sort of science . . . "[31] and every branch of scientific inquiry exhibits "the vital power of self-correction".[32] The reason the sciences are self-corrective is that (in Peirce's view) they utilize methods which are self-corrective. It was thus incumbent on Peirce to show that all the methods of science exhibit self-correction and thereby guarantee progress towards the truth. Those methods for Peirce are threefold: deduction, induction and abduction.

It is at this point that the first of Peirce's serious problem slides occurs. Although he was presumably obliged to show that all three methods of science are self-corrective, he ignores deduction and less excusably, abduction, and limits his discussion almost entirely to induction. There is, nonetheless, a certain rationale for this since, in Peirce's view, inductive methods are operative in every appraisal we make of a theory. So long as the inductive step is self-corrective, any failure of self-correction in deduction and abduction may be ameliorated. Thus, Peirce's problem is changed from that of showing that scientific methods generally are SCMs, to demonstrating that the various methods of induction are self-corrective. The "induction" in question refers to the entire machinery for the testing of a scientific hypothesis. Although the precise significance of the term "induction" undergoes several notorious shifts in his long career, this very general sense of the term is a persistent feature of almost all his discussions of the question.

Thus, in about 1901, Peirce wrote that "the operation of testing a hypothesis by experiment . . . I call *induction*".[33] In 1903 he virtually repeated this definition,[34] and in his later, important essay on "The Varieties and Validity of Induction" (*c*. 1905) he made substantially the same point:

The only sound procedure for induction, whose business consists in testing hypotheses . . . is to receive its suggestions from the hypothesis first, to take up the predictions of experience which it conditionally makes, and then try the experiment . . . When we get to the inductive stage what we are about is finding out how much like the truth our hypothesis is . . .[35]

Peirce asserts on a number of occasions that induction conceived in this broad sense is self-corrective in nature. As early as 1883, he observed that: "We [must not] lose sight of the constant tendency of the inductive method to correct itself. This is of its essence, this is the marvel of it."[36] He reiterated this point twenty years later: "[Induction] is a method of reaching conclusions which, if it be persisted in long enough, will assuredly correct any error concerning future experience into which it may lead us."[37] Between these two temporal extremes, Peirce regularly returns to SCT. About 1896 for instance, he remarked that "Induction is that mode of reasoning which adopts a conclusion as approximative [i.e., approximately true], because it results from a method of inference which must generally lead to the truth in the long run."[38] And two years later he smugly claimed that the fact "that induction tends to correct itself, is obvious enough".[39] To this point, the Peircean texts I have cited could have been written by LeSage, Hartley, Whewell or any of a dozen other methodologists living in the century before Peirce.

What Peirce usually perceived, which his predecessors had not, was that it was not all that obvious that induction, defined as the testing of an hypothesis, is, or tends to be, self-correcting. He saw this as a genuine problem and one which he attempted to resolve on several occasions, most notably in the Lowell Lectures of 1903, and in the famous manuscript "G" (*c*. 1905). In his classic eassy of 1903, Peirce distinguished three varieties of induction: *crude* induction, *qualitative* induction, and *quantitative* induction. *Crude* induction is concerned with universal (as opposed to statistical) hypotheses, the evidential base for which is flimsy and precarious in that they are merely empirical generalizations of the type "all swans are white" or "all Germans drink beer". What typifies crude inductions is not so much the logical form of their conclusions as the nature of the evidential base on which they rest. The only license required for making a crude induction of the form "All A are B" is "the *absence* of [any known] instances to the contrary".[40] Such inductions may be indispensable to daily life but, on Peirce's view, they play no significant role in science. *Quantitative* induction, on the other hand, is an argument from the observed distribution of certain properties in a sample to an hypothesis about the relative distribution of those properties in a larger population. Quantitative induction is induction by simple enumeration in its most literal sense. The conclusion of a quantitative induction is always a statement concerning the probability "that an individual member of a certain experiential class, say the S's, will have a certain character, say that of being P".[41] Unlike crude induction, quantitative induction is (according to Peirce) used in the sciences, if only to a limited extent.

"Of a more general utility" is the remaining variety of induction, *qualitative* induction.[42] This corresponds, more or less, to what is usually called the hypothetico-deductive method. Here, the scientist formulates an hypothesis, deduces predictions from it, and performs experiments to check the predictions. If all of the tested predictions are confirmed, this hypothesis should be tentatively adopted; while if any of the predictions are refuted, the scientist modifies the hypothesis, or abandons it and tries another.

Peirce then argues that one of these species of induction, namely the quantitative variant, is genuinely and demonstrably self-corrective. "Quantitative induction", he insists, "always makes a gradual approach to the truth, though not a uniform approach".[43] Peirce's argument for the self-correcting character of quantitative induction is a crude version of the arguments advanced more recently by Reichenbach and Salmon. Provided that our sampling procedures are fair and that our long run is long enough, the estimates which quantitative inductions lead us to posit will in time approximate ever more closely to the true value.[44] (In developing this argument, Peirce tells us that he was impressed, as Hartley and LeSage had been 150 years earlier, by the fact that "certain methods of mathematical computation correct themselves".)[45]

Ignoring the familiar technical difficulties with this argument,[46] let us concede that Peirce came close to showing that quantitative inductions are self-corrective. At all events, quantitative induction does satisfy two conditions for a self-correcting method; namely, it is a method which not only allows for the refutation of an hypothesis but which also mechanically specifies a technique for finding a replacement for the refuted hypothesis[47] (provided, and it is a crucial proviso, that the hypothesis is taken as a probability statement).

But what of that scientifically more significant species of induction, the method of hypothesis? Such qualitative inductions clearly satisfy the first condition for an SCM, insofar as persistent application of the method of hypothesis will eventually reveal that a false hypothesis is, in fact, false. But the method of qualitative inductions provides no machinery whatever for satisfying the second necessary condition for an SCM; given that an hypothesis has been refuted, qualitative induction specifies no technique for generating an alternative which is (or is likely to be) closer to the truth than the refuted hypothesis. Nor does it even provide a criterion for determining whether an alternative is closer to the truth. Peirce, in short, gives no persuasive arguments to establish that qualitative induction is either strongly or weakly self-corrective.[48]

At a certain level of consciousness, Peirce was fully aware of the fact that he had not shown qualitative inductions to be self-corrective. He remarks that while quantitative induction "always makes a gradual approach to the truth . . . qualitative induction is not so elastic. Usually either this kind of induction confirms the hypothesis or else the facts show that some alteration must be made in the hypothesis."[49] What the facts do not show, of course, is how the hypothesis is to be altered so as to bring it closer to the truth. While "the results of [qualitative] induction *may* help to suggest a better hypothesis", there is no guarantee they will yield a better one.[50]

In one especially candid lecture (1898) on the "Methods for Attaining Truth", Peirce confesses that in "the Explanatory Sciences", we have no sure way of knowing whether the outcome of any confrontation between competing theories is "logical or just".[51] Peirce had evidently landed himself in a situation in which he is pursuing a rapidly degenerating problem. Where before he had answered the question "Are the methods of science self-corrective?" by replying that at least all the inductive methods of science are self-corrective, he is here reduced to saying that even of the various methods of induction, only one is known to be genuinely self-corrective.

Peirce must have sensed the awkwardness of the position in which he found himself. Having set out to show that science is a progressive, self-corrective enterprise, moving ever closer to the truth – and there can be no doubt that this was his initial problem, since both the tradition he was in and his early writings make this clear – Peirce finds himself able to show only that one of the methods of science (and that, by Peirce's admission, a relatively insignificant one) was self-corrective.

I cannot stress too strongly how important it is to be clear about Peirce's intentions. Virtually all Peirce's recent commentators have seen him as setting out to answer Hume's doubts about induction; and have, accordingly, discussed his accounts of SCT and enumerative induction as if they were intended only or primarily as an answer to Hume. Unless my analysis is completely wrong-headed, this is to judge Peirce by an inappropriate yardstick. It was not enumerative induction, but *science* which Peirce set out to justify; it was not Hume but the cynical critics of science whom Peirce set out to answer. (I might generalize this point by adding, parenthetically, that it is one of the wilder travesties of our age that we have allowed the myth to develop that 19th-century philosophers of science were as preoccupied with Hume as we are. As far as I have been able to determine, none of the classic figures of 19th-century methodology – neither Comte, Herschel, Whewell, Bernard, Mill, Jevons, nor Peirce – regarded Hume's arguments about induction as much more than the musings of an historian. This claim is borne out by the fact that in Peirce's thirty-two papers on induction and scientific method – papers teeming with historical references – there is only one reference to Hume; and that is not in connection with the problem of induction but with the problem of miracles.)[52]

As it turned out, Peirce attempted to bridge the gap between intention and performance by a combination of bluster and repetition. Just as LeSage and Hartley could, a century earlier, gloss over their failure to demonstrate an analogy between approximative techniques in mathematics and the methods of science, so Peirce conveniently ignores his painstaking discrimination between the various forms of induction, and pretends (as the quotations above make clear) that his argument has established that all forms of induction (and, by implication, all scientific inferences) are SCMs. In his later writings,[53] he will generally assert that qualitative inductions (or, as he sometimes calls them, 'Inductions of the Second Order') are progressive and self-corrective; but he never goes further than *asserting* that such methods are SCMs, without even the pretense of an argument for that assertion.

Lenz has charitably said that Peirce's "remarks on the self-correcting nature of the broader form[s] of induction are extremely hard to comprehend".[54] I think we must lay a more serious charge at Peirce's feet than that of obscurity. Peirce's remarks in themselves are not difficult to comprehend; he says quite plainly that all forms of induction are self-corrective. What *is* hard to comprehend is Peirce's reason for making such a general assertion. And I think it would be less than candid not to say that Peirce offers no cogent reasons, not even mildly convincing ones, for believing that most inductive methods are self-corrective. I suspect that the explanation of this glaring oversight may be found by recalling Peirce's original motivation.[55]

Peirce began, as I claimed before, with a very general and a very interesting problem: that of justifying scientific inference by showing that the methods of science (including all species of induction) are self-corrective. This was, as

I have shown, one of the standard problems of philosophy of science by Peirce's time. Unable to find a general solution to that problem, Peirce tackles the more limited task of showing that one family of inductive arguments, quantitative inductions, are self-corrective. Having shown, at least to his own satisfaction, that quantitative induction is self-corrective, Peirce then, without even the hint of a compelling argument, makes the crucially serious slide. Seemingly unwilling to admit, even to himself, that he has failed in his original intention to establish SCT for all the methods of science, Peirce acts as if his arguments about quantitative induction show all the other species of induction to be self-corrective as well.

His dilemma was genuine. Having discovered that he could show only quantitative induction to be self-corrective, he could have gone the way of Reichenbach and argued that quantitative induction was the only species of scientific inference, to which all other legitimate methods could be reduced. But Peirce did not share Reichenbach's belief that complex inference was a composite of simple inductions by enumeration. Alternatively, he could have abandoned SCT altogether, conceding that science uses methods which are not, so far as we know, self-corrective. But that would have meant taking much of the flesh out of his philosophy of science. Faced with two such debilitating alternatives, Peirce conveniently ignored the restricted scope of his argument and (perhaps unconsciously) slid from the self-corrective character of the straight rule to SCT as a general thesis. The extent to which Peirce was prepared to make this leap is illustrated by such remarks as his claim that "inquiry of every type, fully carried out, has the vital power of self-correction and growth".[56]

At one point, his bedrock commitment to SCT, even in the absence of any methodological rationale for it, becomes clear:

It is certain that the only hope of retroductive reasoning [viz., qualitative induction] ever reaching the truth is that there may be some natural tendency toward an agreement between the ideas which suggest themselves to the human mind and those which are concerned in the laws of nature.[57]

Unable to find a rational justification for his intuition that science is self-corrective, the otherwise tough-minded Peirce had to fall back on Galileo's *il lume naturale*, on an inarticulate faith in the ability of the mind somehow to ferret out the truth, or a reasonable facsimile thereof:

We shall do better to abandon the whole attempt to learn the truth . . . unless we can *trust* to the human mind's having such a power of guessing right that before many hypotheses shall have been tried, intelligent guessing may be expected to lead us to the one which will support all test . . . .[58]

A similar belief was shared by Peirce's contemporary Pierre Duhem, who argued for SCT in terms of an approach to "*the* natural classification". In a more explicit manner than Peirce, Duhem concedes that he can produce no logically compelling grounds for believing that the history of science brings us closer and closer to a genuine representation of natural relations. Nonetheless, he is convinced that this occurs and that every scientist knows that SCT is true:

Thus, physical theory never gives us the explanation of experimental laws . . . but the more complete it becomes the more we apprehend that the logical order in which theory orders experimental laws is the reflection of an ontological order, the more we suspect that the relations it establishes among the data of observation correspond to real relations among things . . . . The physicist cannot take account of this conviction . . . . But while the physicist is powerless to rid his reason of it . . . yielding to an intuition which Pascal would have recognized as one of those reasons of the heart "that reason does not know", he asserts his faith in a real order reflected in his theories more clearly and more faithfully as time goes on.[59]

Less optimistic than Peirce about the possibility of finding a methodological rationale for the view that science moves ever closer to the truth, Duhem maintains that the methodologist cannot justify SCT, and that its only defense lay in what Duhem calls a "metaphysical assertion".[60]

To return to Peirce only briefly, I suspect that there is another important sense in which he takes much of the force out of the SCT tradition. As I have tried to make clear, that tradition was committed to the view (among others) that the replacement of one non-statistical hypothesis by another was the basic unit of progress and self-correction. Peirce, at least on some occasions, abandons that view altogether. In its place, he argues that, although we have no way of correcting our hypotheses, what we can correct are the assignments of probability which we give to those hypotheses. When arguing in this vein, Peirce sees the process of assigning probabilities to hypotheses as self-corrective, while the process of replacing one hypothesis by another no longer remains even a candidate for consideration as a self-corrective process. In the course of time, it is not our *theories* which get closer to the truth, but rather, the *probabilities* which we assign to theories exhibit progress and self-correction. Where all previous discussions of the question had been concerned to show that a sequence of hypotheses of the form:

A is B,
A is C,
A is D,
etc.,

is progressive and self-corrective, Peirce's quantitative induction goes for the "cheapest" form of self-correction, arguing that a sequence of the following kind:

The probability that *A* is *B* is m/n,
The probability that *A* is *B* is m'/n',
The probability that *A* is *B* is m''/n'',
etc.,

is (or can be) self-corrective. Peirce simply cannot handle a case where an hypothesis (of the form "*A* is *B*") is replaced by a conceptually different one (say "*A* is *C*").[61]

It would not be appropriate in this volume to discuss at length the views of more recent methodologists about SCT, since that would take us well into the 20th century. Nonetheless, I think a few words are in order about more recent developments in so far as they link up rather closely to the tradition I have been discussing here. As everyone knows, Hans Reichenbach took up SCT, most notable in his *Wahrscheinlichkeitslehre* and his *Experience and Prediction*. In both works, Reichenbach, like Peirce, set out to show that the straight rule is self-corrective, that induction by simple enumeration is a SCM. Like Peirce, Reichenbach then goes on to assume, with only the flimsiest of arguments, that science is a self-correcting enterprise because (and here Reichenbach differs from Peirce) all the methods of science can be reduced to enumerative induction.[62] Unfortunately, however, Reichenbach's attempts to reduce most scientific methods to convoluted species of the straight rule are at best, programmatic; at worst, unconvincing. As a result Reichenbach, like Peirce, found himself unable to prove SCT generally, and was forced to be content with the comparatively insignificant consolation that enumerative induction is self-corrective.

All the same, it must be said on Reichenbach's behalf that he takes up the banner of the SCT tradition in a less half-hearted way than Peirce had. Reichenbach sensed the object of the exercise, and understood that exploration of the self-correcting properties of the straight rule was only of crucial import in so far as one could establish the relevance of the straight rule to more subtle forms of scientific reasoning. That Reichenbach's program did not come off, that he never quite managed to achieve the reduction of scientific methods to enumerative induction, does not diminish the soundness of his intuitions as to the nature of the problem.

In the last two decades, however, there seems to have been a tendency to return to a Peircean rather than a Reichenbachian treatment of the question.

Many contemporary philosophers of science, perhaps forgetting that self-correction was originally a thesis about science rather than a putative answer to Hume, have explored at length the question whether the method of enumerative induction is self-corrective without seriously considering whether the methods of science are enumerative.[63] Reichenbach's most distinguished disciple, Wesley Salmon, similarly skirts this particular issue on many occasions.[64] One has the impression (perhaps unjustifiably) that such philosophers have become so involved with the technical and formal aspects of Peirce's solution that they have lost sight of the problem to which it was a solution. We are, I suspect, sometimes repeating Peirce's mistake of thinking that so long as we establish that *any* ampliative inference is self-corrective, we can easily show that most of them are.

Criticisms of the type I have offered here, however well intentioned, are always open to the charge of being premature and philistinistic. After all, one might say, a break-through could come at any moment and in that event the work of Reichenbach and Salmon might become to the foundations of scientific inference what Russell, Frege or Cantor were to the foundations of mathematics. Moreover, it might be pointed out that foundational studies, especially in their preliminary stages, always have only tenuous connections to what they purport to be foundations of. But, granting all that, one has a right to insist that putative "foundational studies" must satisfy some canons of adequacy, and be subject to certain standards of criticism.

Precisely what these standards are I do not pretend to know. (This in itself is a major philosophical problem.) But there are several seemingly relevant points to make about the so-called pragmatic justification of induction and scientific inference. The first point is that distinguished philosophers have been exploring this approach for almost a century. In that time, they are no closer to exhibiting a connection between the straight rule and other modes of inference than Peirce was in 1872. While promissory notes are not dated, there is a presumption that payments will be made at respectable intervals. Secondly, and more disturbingly, the 'pragmatic' approach has, at least since the 1930s, tended to concern itself less and less with the one thesis which originally made that approach interesting, viz., the thesis that scientific inference could be reduced to enumerative inference. The centrality of that thesis in the pragmatic tradition has been replaced by a preoccupation with enumerative induction itself. In an unnoticed sleight of hand, *the problem of the justification of science has been displaced by the problem of justifying induction*. And, in the absence of any established link between the former and the latter, this portion of the "philosophy of science" has surrendered any convincing claims to being the philosophy of science.

If we believe, with Peirce, LeSage, Hartley, Whewell, and Duhem that science is a self-corrective, progressive enterprise, then we should presumably be seeking to show how and why it is so. If we further believe with Peirce and Reichenbach that the exploration of enumerative induction will provide the answer, then we ought to be exploring more assiduously the role of enumerative induction in real science. What we must avoid is falling into the Peircean pit by assuming without argument that the grand old problem of the progress of science is apt to be clarified by technical investigations of the straight rule. We have accepted Peirce's *ersatz* self-correction – a self-correction which only can deal with changes in probabilities rather than changes in theories – without openly discussing whether full-bodied self-correction in the traditional sense is beyond our powers of explication. It is the self-correcting nature of science, not the self-corrective nature of a "puerile" rule, which should be our main concern.[65]

NOTES

* Since this chapter first appeared, it has been discussed at length by Nicholas Rescher and Ilkka Niiniluoto. It should be evaluated in the light of the constructive criticisms they have made of its central thesis.

1 *Mid-West Quarterly* 2 (1914), 49.

2 There is a vast body of exegetical and critical literature dealing with Peirce's philosophy. The following are concerned explicitly with Peirce's treatment of self-correction: A. W. Burks, 'Peirce's Theory of Abduction', *Philosophy of Science* 13 (1946), 301–6. C. W. Cheng, *Peirce's and Lewis's Theories of Induction* (The Hague, 1969), H. G. Frankfurt, 'Peirce's Notion of Abduction', *Journal of Philosophy* 55 (1958), 593–7; J. Lenz, 'Induction as Self-Corrective', in Moore and Robin (eds.), *Studies in the Philosophy of Charles Sanders Peirce* (Amherst, Mass., 1964); E. Madden, 'Peirce on Probability', in ibid.; F. E. Reilly, *The Method of the Sciences According to C. S. Peirce*, Doctoral dissertation, St Louis University, 1959. While acknowledging a debt to all of these works, I believe it is fair to say that none of these authors treats Peirce's approach to SCT within the historical framework in which I have tried to place it.

3 See, for instance, Burks, 'Peirce's Theory'. Even Peirce himself tries to give the impression that he was the first to enunciate the view that scientific reasoning is self-corrective. For instance, he wrote in 1893 that "you will search in vain for any mention in any book I can think of" of the view "that reasoning tends to correct itself". C. S. Peirce, *Collected Papers*, ed. Hartshorne, Weiss et al., 8 vols. (Cambridge: Harvard University Press, 1931–58) Vol. 5, p. 579. Without questioning Peirce's integrity, we do have some grounds for doubting his memory. Peirce makes numerous references to the works of may of the writers whom I cite below as Peirce's predecessors in this matter. (See, for

example, ibid., Vol. 5, p. 276 n., where he writes knowledgeably of the philosophies of science of both LeSage and Hartley, who had stressed the self-correcting aspects of scientific reasoning.)

[4] This point requires some qualification. As is well known, passages can be adduced from all these authors where they seem to abandon the infallibilism of TICT and to replace it by a more modest "probabilism". (Many of the relevant texts are discussed in Chapter 4.) However, it would be a serious error of judgment to let these concessions to fallibilism obscure the fact that all of these figures shared the classical view that science at its best is *demonstrated knowledge from true principles*. Bacon, Descartes, Locke, and Boyle all see it as a goal that science become infallible; until that goal is realized they are willing to settle – but only temporarily – for merely probable belief. Their long-term aim, however, is to replace such mere opinion by genuine knowledge.

[5] See Robert Hooke's posthumously published account of "inductive logic" in *The Posthumous Works of Robert Hooke*, ed. R. Waller (London, 1705), pp. 3 ff.

[6] See Chapter 4.

[7] A century and a half later Max Planck gave eloquent expression to this quintessentially 18th-century viewpoint: "Nicht der Besitz der Wahrheit, sondern dass erfolgreiche Suchen nach ihr befruchtet und beglüchte den Froscher". (*Wege zur physikalischen Erkenntnis*, 4th ed. [Leipzig, 1944], p. 208.)

[8] Some, but by no means all. As late as the 1790s, philosophers such as Thomas Reid were still arguing for a strictly inductive methodology. (Cf. Chapter 7.)

[9] To be faithful to the historical situation, it is important to point out that some 18th- and early 19th-century methodologists, while accepting SCT as a general thesis, were not altogether happy with the idea expressed in (2) above. As formulated there, SCT is committed to the view that there is a *mechanical* process for finding alternatives. Some methodologists denied this. What they did insist on, however, was that: (2′) Science possess techniques for determining unambiguously whether an alternative T' is closer to the truth than a refuted T.

William Whewell, for instance, denied the claim implicit in (2) that the scientist possessed any algorithm for automatically correcting an hypothesis. Nonetheless, he was convinced that it was generally possible, given a (refuted) theory and an alternative to it, to determine which of the two was (in Whewell's language) "nearer to the truth". Hereafter, I shall refer to the pair (1) and (2) as the *strong* self-correction thesis (or SSCT) and to the pair (1) and (2′) as the *weak* thesis of self-correction (WSCT).

There is another important qualification to make here. Although all the figures I discuss talk about "getting closer to the truth", "moving nearer to the truth", etc., it is not altogether clear that there is a shared conception of what truth consists in. With some writers, for instance, the notion of truth seems to be an instrumental one (viz., that is true which adequately "saves the phenomena"); with others, the concept of truth is a correspondence one. Nonetheless, most discussions of self-correction and proximity to the truth seem to be conducted independently of various conceptions of, and criteria for, the truth.

[10] See note 9 above.

[11] This claim for the priority of LeSage and Hartley is, like all claims for historical priority, necessarily tentative. R. V. Sampson, in his *Progress in the Age of Reason* (London, 1956), asserts that Blaise Pascal conceived of science as "cumulative, self-corrective" and progressive. I have been unable to find such an argument in Pascal and

(unfortunately) Sampson offers no evidence for his interpretation. Similarly, Charles Frankel (*The Faith of Reason*, [New York, 1948]) argues likewise without evidence, that "For Pascal . . . scientific method was progressive because it was public, cumulative, and self-corrective" (p. 35). Until more substantive evidence is produced, I believe the available historical evidence supports my priority claims for Hartley and LeSage. However, the argument in the body of the essay does not depend on the priority issue.

12 David Hartley, *Observations on Man* (London, 1749), 1: 341–2.

13 Ibid., 1: 345–6. Basically, the rule of false position worked as follows: If one sought the solution to an equation of the form $ax + b = 0$, one made a conjecture, m, as to the value of x. The result, n, of substituting m for x in the left-hand side of the equation is given by $am + b$. The correct value of x was then determined by the formula

$$x = mb/(b-n)$$

The rule of false position was one of the earliest known rules for the solution of simple equations.

It should be added that during the 18th century, the term "rule of false position" normally referred, not specifically to the rule given above, but rather to what we call the rule of double position, which involves two conjectures rather than one. An interesting discussion of this latter rule may be found in Robert Hooke's *Philosophical Experiments and Observations*, ed. Durham (London, 1726), pp. 84–6.

14 R. Recorde, *Ground of Artes* (London, 1558), fol. Z4.

15 *Observations on Man*, n12, 1: 349.

16 Ibid., 1: 349. It is perhaps worth observing that Hartley adhered to a theory of moral progress and self-improvement which paralleled the progress and self-correction of science. "We have", he writes, "a Power of suiting our Frame of Mind to our Circumstances, of correcting what is amiss, and improving what is right" (Ibid., 1: 463).

17 Ibid., 1: 349. There is, we should observe, a very great difference in the results which these various "approximative methods" yield. Some of these methods – such as the Newtonian method of approximation to the roots of a general equation – do not necessarily ever yield a true result. We can, by their use, constantly improve our estimate, but there is no guarantee that we will ever determine precisely the correct answer. However, other methods Hartley mentions, especially the rule of false position, not only correct a false guess, but immediately replace it by the correct solution. These differences become very significant when applied to a scientific context. If our model for scientific method is the rule of false position, then one can imagine science rapidly reaching a stage where all the false theories have been replaced by true ones, and where scientific knowledge would be both static and non-conjectural. If, on the other hand, our model for inquiry is the search by approximation for the roots of an equation, then science would seem to be perhaps perennially in a state of change and flux, with no guarantee whatever that it could ever reach the final truth.

Hartley, as well as most of his 19th-century successors, seems to vacillate between these two very different models.

18 Talk of a 'self-correcting method' is, of course, slightly misleading since the method does not correct itself, but rather it allegedly corrects those statements which an earlier application of the method produced. However, since linguistic traditions sanctify all manner of confusions, and since it is de rigueur to speak of methods with these properties as self-corrective methods, I will do so, hoping the reader will bear this *caveat* in mind.

[19] A method will be weakly self-corrective (WSCM) if (a) above and if (b) without itself specifying a 'truer' alternative, it can determine for certain whether a given alternative is truer. (See also Note 6 above).

[20] Precisely this criticism was raised by Condillac in 1749 against the view that science can borrow the approximative methods of the mathematician. (Cf. his *Traité des Systèmes* [Paris, 749], pp. 329–31.) It was also raised by J. Senebier a generation later. (Cf. his *Essai sur l'Art d'Observer et de Faire des Expériences* [Génève, 1802], 2: 215–6.)

[21] As LeSage puts it: "The corrections made of these particular suppositions, resulting from the small multiplications which serve to test their validity, have as their sole aim to bring closer together these suppositions and the [true] number; with the exception of the last partial division, which must be performed rigorously because it is here that one finally rejects the inaccuracies one has permitted oneself in the previous operations." G. H. LeSage, 'Quelques Opuscules rélatifs a la Méthode', posthumously published by Pierre Prevost in his *Essais de Philosophie* (Paris, 1804), 2: 253–35. The passage in question dates from the 1750s, and appears on p. 261. (I discuss LeSage's work at much greater length in Chapter 8.)

[22] Ibid., 2: 261.

[23] Joseph Priestley, *The History and Present State of Electricity* (London, 1767), p. 381.

[24] "Souvent on s'écarte du vrai, sans douter, et on le fuit en croyant le poursuivre" (*Essai*, 2: 220).

[25] Prevost, *Essais de Philosophie* (Paris, 1804), 2: 196. Prevost nevertheless believes that there are self-corrective methods which the scientist can use.

[26] See, for instance, the several essays on progress in Whewell's *Philosophy of Discovery* (London, 1860) and Auguste Comte's preliminary discourse to the *System of Positive Polity* (4 vols. [London, 1875–7]). Similar, if more vague, sentiments are involved in John Herschel's discussion (*Preliminary Discourse on the Study of Natural Philosophy* [London, 1831], para. 224 ff.).

[27] E. Renan, 'Claude Bernard', in Renan (ed.), *L'Oeuvre de Claude Bernard* (Paris, 1881), p. 33. My italics.

[28] T. H. Huxley, *Hume* (London, 1894), p. 65.

[29] My labels are, of course, anachronistic. The concepts they denote are not.

[30] For Peirce's application of SCT to the history of science, see his *Lessons from the History of Science*, (c. 1896), in *Collected Papers*, 1: 19–49, especially para. 108, p. 44.

[31] *Collected Papers*, 5: 579.

[32] Ibid., 5: 582.

[33] Ibid., 6: 526; cf. also 2: 755.

[34] Ibid., 7: 110.

[35] Ibid., 2: 775.

[36] Ibid., 2: 729.

[37] Ibid., 2: 769.

[38] Ibid., 1: 67.

[39] Ibid., 5: 576. Other relevant passages would include: 1868 (revised 1893): "we cannot say that the generality of inductions are true, but only that in the long run they approximate to the truth" (5: 350). 1898: "A properly conducted inductive research corrects its own premises" (5: 576). 1901: "[Induction] commences a proceeding which

must in the long run approximate to the truth" (2: 780). "persistently applied to the problem [induction] must produce a convergence (through irregular) to the truth" (2: 775). "the method of induction must generally approximate to the truth" (6: 100). 1903: "The justification of [induction] is that, although the conclusion at any stage of the investigation may be more or less erroneous, yet the further application of the same method must correct the error" (5: 145). "Suppose we define Inductive reasoning as that reasoning whose conclusion is justified . . . by its being the result of a method which if steadfastly persisted in must bring the reasoner to the truth of the matter or must cause his conclusion in its changes to converge to the truth as its limit" (7: 110). ". . . if this mode of reasoning [viz., induction] leads us away from the truth, yet steadily pursued, it will lead to the truth at last" (7: 111). See also *Collected Papers*, 2: 709.

40 Ibid., 2: 756.

41 Ibid., 2: 758.

42 Ibid., 2: 77 ff.

43 Ibid., 2: 770.

44 Provided, of course, that there *is* some limit to the sequence in question; a qualification which Peirce realized to be essential.

45 Ibid., 5: 574. Peirce's example, that of the extraction of roots, is identical to Hartley's and LeSage's. It is perhaps appropriate to add here that Peirce knew Hartley's *Observations on Man* first-hand, and makes numerous references to it in his *Collected Papers*. Moreover, he knew of LeSage's work, at least second-hand, citing it in volume 5 of his *Collected Papers*.

I know too little about Peirce's intellectual biography to assert with any confidence that it was definitely Hartley and LeSage who gave him the idea of a SCM; but, given Peirce's knowledge of Hartley and the obvious similarities in the initial approaches to the problem, it seems a reasonable conjecture that Hartley may have stirred Peirce to consider the question of self-correction in detail.

46 For references to the vast body of technical literature on the straight rule, cf. the bibliography in Salmon's *The Foundations of Scientific Inference* (Pittsburgh, 1967).

47 Whether that replacement is closer to the truth than that which it replaces, is, of course, another matter. But at least quantitative induction can specify a replacement, and is thus (potentially) a strong self-corrective method.

48 This point, viz., that qualitative induction is not (or, at least, has not been shown to be) self-corrective, has gone unnoticed by several of Peirce's commentators. For instance, Cheng writes: "To say that a qualitative induction is self-correcting is either to say that a given hypothesis is replaceable by a new hypothesis or that the scope of the given hypothesis is modifiable or limitable . . . " (*Peirce's and Lewis's Theories*, p. 73).

In arguing this point, Cheng has used an unfortunate sense of 'self-correcting'. That an hypothesis is replaceable or 'modifiable' merely means that we have techniques for discarding or altering it. If qualitative induction is to be self-correcting then we need, at a minimum, the further assurance that its replacement or altered expression is an improvement. This assurance Peirce nowhere provides, and on occasion even denies that we can obtain it.

49 *Collected Papers*, 2: 771.

50 Ibid., 2: 759.

51 Ibid., 5: 578.

52 The titles of these 32 papers are listed in appendix I to Cheng's *Peirce's and Lewis's*

Theories of Induction. Ironically, Cheng himself discusses Peirce's work as if it were designed explicitly as a reply to Hume.

53 *Collected Papers*, 7: 114–9.

54 J. W. Lenz, 'Induction as Self-Corrective', in E. Moore and R. Robin (eds.), *Studies of Peirce*, n. 2, p. 152. Cheng echoes Lenz when he observes that "Peirce does not make clear what the self-correcting process of induction means ... " (*Peirce and Lewis's Theories of Induction*, p. 67).

55 One could schematically survey the major changes in SCT by looking at three formulations, the first, typically 18th-century, the second, 19th-century, and the third, Peirce's:

SCT_1: The methods of science are such that, given a refuted hypothesis *H*, a mechanical procedure exists for generating a 'truer' *H'* .... Science is progressive (i.e., getting closer to the truth).

SCT_2: The methods of science are such that, given a refuted hypothesis *H*, we can always determine whether an alternative *H'* is 'truer' .... Science is progressive.

SCT_3: The method of enumerative induction is such that, given a refuted *H* (and the available evidence) we can mechanically produce an alternative *H'* which is likely to be truer than *H* .... Science is progressive.

The sequence SCT_1–SCT_2–SCT_3 is one in which the premises become increasingly preicse and defensible; but the price paid is that the premises seem to lend less and less inferential support to the conclusion.

56 *Collected Papers*, 5: 582.

57 Ibid., 1: 81. Peirce insists "that it is a primary hypothesis ... that the human mind is akin to the truth in the sense that in a finite number of guesses it will light upon the correct hypothesis" (Ibid., 7: 220).

58 Ibid., 6: 531. Cf. also 1: 121.

59 Pierre Duhem, *The Aim and Structure of Physical Theory*, trans. Wiener (New York, 1962), pp. 26–7.

60 Ibid., p. 297. Duhem summarizes his position when he observes: "To the extent that physical theory makes progress, it becomes more and more similar to a natural classification which is its ideal end. Physical theory is powerless to prove this assertion is warranted, but if it were not, the tendency which directs the development of physics would remain incomprehensible. Thus, in order to find the title to establish its legitimacy [as an SCM], physical theory has to demand it of metaphysics" (Ibid., p. 298).

61 My suspicion is that this 'cheap' form of inductive self-correction has its origins in Laplace's rule of succession, and the discussions that rule engendered in 19th-century probability theory.

62 Reichenbach writes: "The method of scientific inquiry may be considered as a concatenation of [enumerative] inductive inference ... " (*Experience and Prediction*, [Chicago, 1938], p. 364).

63 Cf. G. H. von Wright, *The Logical Problem of Induction*, 2nd ed. (Oxford, 1965), chap. viii. It is, however, to von Wright's credit that he, almost alone among Peirce's commentators, perceives the limited scope of Peirce's treatment of self-correction. As he puts the point: "the Peircean idea of induction as a self-correcting approximation to the truth has no immediate significance ... for other types of inductive reasoning than statistical generalization" (Ibid., p. 226).

[64] See, for instance, W. Salmon, 'Vindication of Induction', in H. Feigl and G. Maxwell (eds.), *Current Issues in the Philosophy of Science* (New York, 1961), p. 256; and W. Salmon, 'Inductive Inference', in B. Brody (ed.), *Readings in the Philosophy of Science* (Englewood Cliffs, N. J., 1970), p. 615.

[65] A very different formulation of SCT has been developed by Karl Popper in his *Conjectures and Refutations* (London, 1963). Popper's approach, unlike that of Peirce, Reichenbach and Salmon, does not attempt to make enumerative induction the cornerstone of scientific inference. It depends, rather, upon showing (unsuccessfully, I believe) that the method of hypothesis is weakly self-corrective in virtue of methodological conventions about increases in content. Popper is perhaps alone among contemporary philosophers of science in facing the issues raised by SCT in their full generality. As inadequate as his discussion of verisimilitude is, he has sensed the magnitude of the problem. In this, as in other ways, Popper is probably closer to the 19th-century methodological tradition than is any other living philosopher.

BIBLIOGRAPHIC NOTE

The individual chapters of this book contain references to all the works specifically cited. There is, of course, a vast body of scholarly studies on the history of methodology which I have not included. A guide to much of that literature can be found in L. Laudan, 'Theories of Scentific Method from Plato to Mach: A Bibliographic Review', *History of Science* 7 (1968), 1–63.

INDEX OF NAMES

THE UNIVERSITY OF WESTERN ONTARIO SERIES IN PHILOSOPHY OF SCIENCE

A Series of Books in Philosophy of Science, Methodology, Epistemology, Logic, History of Science, and Related Fields

1. J. Leach, R. Butts, and G. Pearce (eds.), *Science, Decision and Value*. Proceedings of the Fifth University of Western Ontario Philosophy Colloquium, 1969. 1973, vii + 213 pp.
2. C. A. Hooker (ed.), *Contemporary Research in the Foundations and Philosophy of Quantum Theory*. Proceedings of a Conference held at the University of Western Ontario, London, Canada, 1973. xx + 385 pp.
3. J. Bub, *The Interpretation of Quantum Mechanics*. 1974, ix + 155 pp.
4. D. Hockney, W. Harper, and B. Freed (eds.), *Contemporary Research in Philosophical Logic and Linguistic Semantics*. Proceedings of a Conference held at the University of Western Ontario, London, Canada. 1975, vii + 332 pp.
5. C. A. Hooker (ed.), *The Logico-Algebraic Approach to Quantum Mechanics*. 1975, xv + 607 pp.
6. W. L. Harper and C. A. Hooker (eds.), *Foundations of Probability Theory, Statistical Inference, and Statistical Theories of Science*, 3 Volumes. Vol. I: *Foundations and Philosophy of Epistemic Applications of Probability Theory*. 1976, xi + 308 pp. Vol. II: *Foundations and Philosophy of Statistical Inference*. 1976, xi + 455 pp. Vol. III: *Foundations and Philosophy of Statistical Theories in the Physical Sciences*. 1976, xii + 241 pp.

8. J. M. Nicholas (ed.), *Images, Perception, and Knowledge.* Papers deriving from and related to the Philosophy of Science Workshop at Ontario, Canada, May 1974. 1977, ix+309 pp.
9. R. E. Butts and J. Hintikka (eds.), *Logic, Foundations of Mathematics, and Computability Theory.* Part One of the Proceedings of the Fifth International Congress of Logic, Methodology and Philosophy of Science, London, Ontario, Canada, 1975. 1977, x+406 pp.
10. R. E. Butts and J. Hintikka (eds.), *Foundational Problems in the Special Sciences.* Part Two of the Proceedings of the Fifth International Congress of Logic, Methodology and Philosophy of Science, London, Ontario, Canada, 1975. 1977, x+427 pp.
11. R. E. Butts and J. Hintikka (eds.), *Basic Problems in Methodology and Linguistics.* Part Three of the Proceedings of the Fifth International Congress of Logic, Methodology and Philosophy of Science, London, Ontario, Canada, 1975. 1977, x+321 pp.
12. R. E. Butts and J. Hintikka (eds.), *Historical and Philosophical Dimensions of Logic, Methodology and Philosophy of Science.* Part Four of the Proceedings of the Fifth International Congress of Logic, Methodology and Philosophy of Science, London, Ontario, Canada, 1975. 1977, x+336 pp.
13. C. A. Hooker (ed.), *Foundations and Applications of Decision Theory,* 2 volumes. Vol. I: *Theoretical Foundations.* 1978, xxiii+442 pp. Vol. II: *Epistemic and Social Applications.* 1978, xxiii+206 pp.
14. R. E. Butts and J. C. Pitt (eds.), *New Perspectives on Galileo.* Papers deriving from and related to a workshop on Galileo held at Virginia Polytechnic Institute and State University, 1975. 1978, xvi + 262 pp.
15. W. L. Harper, R. Stalnaker, and G. Pearce (eds.), *Ifs. Conditionals, Belief, Decision, Chance, and Time.* 1980, ix + 345 pp.
16. J. C. Pitt (ed.), *Philosophy in Economics.* Papers deriving from and related to a workshop on Testability and Explanation in Economics held at Virginia Poly-Technic Institute and State University, 1979. 1981.
17. Michael Ruse, *Is Science Sexist?* 1981, xix + 299 pp.
18. Nicholas Rescher, *Leibniz's Metaphysics of Nature.* 1981, xiv + 126 pp.

GPSR Compliance

The European Union's (EU) General Product Safety Regulation (GPSR) is a set of rules that requires consumer products to be safe and our obligations to ensure this.

If you have any concerns about our products, you can contact us on ProductSafety@springernature.com

In case Publisher is established outside the EU, the EU authorized representative is:

Springer Nature Customer Service Center GmbH
Europaplatz 3
69115 Heidelberg, Germany

Batch number: 10823811

Printed by Printforce, the Netherlands